BIOTECHNOLOGY IN AGRICULTURE, INDUSTRY AND MEDICINE

SYNTHETIC BIOLOGY

PROSPECTS AND IMPLICATIONS

BIOTECHNOLOGY IN AGRICULTURE, INDUSTRY AND MEDICINE

Additional books in this series can be found on Nova's website under the Series tab.

Additional E-books in this series can be found on Nova's website under the e-books tab.

LABORATORY AND CLINICAL RESEARCH

Additional books in this series can be found on Nova's website under the Series tab.

Additional E-books in this series can be found on Nova's website under the e-books tab.

BIOTECHNOLOGY IN AGRICULTURE, INDUSTRY AND MEDICINE

SYNTHETIC BIOLOGY

PROSPECTS AND IMPLICATIONS

ROGER E. HAMILTON
AND
LAURA M. MORALES
EDITORS

nova publishers
New York

LIBRARY OF CONGRESS CATALOGING-IN-PUBLICATION DATA

ISBN: 978-1-62257-207-6

Published by Nova Science Publishers, Inc. † New York

CONTENTS

PREFACE

The 21st century is widely heralded as the century of biology. Building on the fundamental understanding achieved in the second half of the last century, revolutionary advances are expected to improve many aspects of our lives, from clean energy and targeted, safer medicines to new industries. Prominent among emerging technologies is "synthetic biology," which aims to apply standardized engineering techniques to biology and thereby create organisms or biological systems with novel or specialized functions to address countless needs. This book examines the Presidential Commission for the Study of Bioethical Issues review on the developing field of synthetic biology which researches and identifies appropriate ethical boundaries to maximize public benefits and minimize risks.

Chapter 1 - PCSBI approached this task through inclusive and deliberative engagement with ethicists, scientists, engineers, and individuals in faith, business, and non-profit communities. We held three public meetings, both in and outside of Washington, D.C., created an open forum for dialogue, and heard many diverse voices. The Commission found that synthetic biology offers extraordinary promise to create new products for clean energy, pollution control, and medicine, to revolutionize chemical production and manufacturing, and to create new economic opportunities. With this promise comes a duty to attend carefully to potential risks, be responsible stewards, and consider thoughtfully the implications for humans, other species, nature, and the environment. PCSBI concluded that synthetic biology is capable of significant but limited achievements posing limited risks. Future developments may raise further objections, but the Commission found no reason to endorse additional federal regulations or a moratorium on work in this field at this time. Instead, the Commission urges monitoring and dialogue between the private and public sectors to achieve open communication and cooperation. The Commission recommends that the government, through a coordinated process or body within the Executive Office of the President, lead an ongoing review of developments, risks, opportunities, and oversight as this field grows. This review should be in consultation with relevant scientific, academic, international, and public communities, and whenever possible its results should be made public. We also recommend that reasonable risk assessment should precede any field release of synthetic organisms. We suggest support for public engagement, education, and dialogue to ensure public trust and avoid unnecessary limitations on science and social progress. You gave the Commission a rare and exceptional opportunity to be proactive and forward looking in this first study. The Commission is grateful for the opportunity to serve you and the nation in this way. We would be happy to brief you if you have any questions about our recommendations.

Chapter 2 - The term "synthetic biology" is used and defined in a variety of ways within the scientific community. Indeed, the dynamic nature of science all but ensures that a precise definition of "synthetic biology" will remain elusive and will evolve over time. The power of synthetic biology, as more broadly envisioned, reflects the possibility of synthetic biological systems that are programmable, self-referential, and modular. The ability to reformat the modules (e.g., the use of altered genetic codes or unnatural amino acids) leads to increased information content and vastly expanded possibilities for new function. Functionally, two disparate experimental approaches can be used to describe synthetic biology: "top down" and "bottom up." The goal of both is to create novel biological structures with predictable properties and functions. The top down approach is related to classical recombinant DNA approaches and involves the re-engineering of existing organisms or genomes for a defined purpose. Increasingly powerful methods for DNA synthesis and assembly have significantly accelerated re-engineering capabilities, however. The bottom up approach entails the assembly of biological components in a variety of novel ways. This approach attempts to assemble systems (both living and nonliving) that perform desired functions in a predictable manner. Additionally, synthetic biology can be viewed even more broadly to include several emerging areas of study (e.g., nanotechnology, biocomputation), and we note that the biosecurity risks and dual use potential of new technologies and scientific information, regardless of their fields of origin, will remain important considerations for all areas of science.

Chapter 3 - This is a statement of J. Craig Venter, President, J. Craig Venter Institute.

Chapter 4 - This is a testimony of Dr. Jay Keasling, University of California at Berkeley.

Chapter 5 - This is a testimony of Dr. Drew Endy, Stanford University.

Chapter 6 - This is a testimony of Gregory E. Kaebnick, The Hastings Center.

Chapter 7 - This is a testimony of Dr. Anthony S. Fauci, Director, National Institute of Allergy and Infectious Diseases, National Institutes of Health.

In: Synthetic Biology
Editors: Roger E. Hamilton and Laura M. Morales
ISBN: 978-1-62257-207-6

Chapter 1

NEW DIRECTIONS: THE ETHICS OF SYNTHETIC BIOLOGY AND EMERGING TECHNOLOGIES*

Presidential Commission for the Study of Bioethical Issues

ABOUT THE PRESIDENTIAL COMMISSION FOR THE STUDY OF BIOETHICAL ISSUES

The Presidential Commission for the Study of Bioethical Issues (PCSBI) is an advisory panel of the nation's leaders in medicine, science, ethics, religion, law, and engineering.

PCSBI advises the President on bioethical issues arising from advances in biomedicine and related areas of science and technology. The Commission seeks to identify and promote policies and practices that ensure scientific research, health care delivery, and technological innovation are conducted in a socially and ethically responsible manner.

For more information about PCSBI, please see www.bioethics.gov.

PRESIDENTIAL COMMISSION FOR THE STUDY OF BIOETHICAL ISSUES

President Barack Obama
The White House;
1600 Pennsylvania Avenue, NW
Washington, DC 20500
Dear Mr. President:

We are pleased to present to you this report, *New Directions: The Ethics of Synthetic Biology and Emerging Technologies*. In response to your request of May 20, 2010, this first report of the Presidential Commission for the Study of Bioethical Issues (PCSBI) examines

* This is an edited, reformatted and augmented version of Presidential Commission for the Study of Bioethical Issues, dated December 2010.

the implications of the emerging science of synthetic biology, including the announcement in May of the successful creation of a self-replicating bacterial cell with a completely synthetically-replicated genome. It offers recommendations to ensure that America reaps the benefits of this developing field within appropriate ethical boundaries.

PCSBI approached this task through inclusive and deliberative engagement with ethicists, scientists, engineers, and individuals in faith, business, and non-profit communities. We held three public meetings, both in and outside of Washington, D.C., created an open forum for dialogue, and heard many diverse voices. The Commission found that synthetic biology offers extraordinary promise to create new products for clean energy, pollution control, and medicine, to revolutionize chemical production and manufacturing, and to create new economic opportunities. With this promise comes a duty to attend carefully to potential risks, be responsible stewards, and consider thoughtfully the implications for humans, other species, nature, and the environment. PCSBI concluded that synthetic biology is capable of significant but limited achievements posing limited risks. Future developments may raise further objections, but the Commission found no reason to endorse additional federal regulations or a moratorium on work in this field at this time. Instead, the Commission urges monitoring and dialogue between the private and public sectors to achieve open communication and cooperation. The Commission recommends that the government, through a coordinated process or body within the Executive Office of the President, lead an ongoing review of developments, risks, opportunities, and oversight as this field grows. This review should be in consultation with relevant scientific, academic, international, and public communities, and whenever possible its results should be made public. We also recommend that reasonable risk assessment should precede any field release of synthetic organisms. We suggest support for public engagement, education, and dialogue to ensure public trust and avoid unnecessary limitations on science and social progress. You gave the Commission a rare and exceptional opportunity to be proactive and forward looking in this first study. The Commission is grateful for the opportunity to serve you and the nation in this way. We would be happy to brief you if you have any questions about our recommendations.

Sincerely,

Amy Gutmann, Ph.D. Chair

James Wagner, Ph.D. Vice-Chair

THE WHITE HOUSE
WASHINGTON

May 20, 2010
Dr. Amy Gutmann
President and Christopher H. Browne
Distinguished Professor of Political Science
University of Pennsylvania
1 College Hall, Room 100
Philadelphia, Pennsylvania 19104-6380

Dear Dr. Gutmann,

As you know, scientists have announced a milestone in the emerging field of cellular and genetic research known as synthetic biology. While scientists have used DNA to develop genetically modified cells for many years, for the first time, all of the natural genetic material in a bacterial cell has been replaced with a synthetic set of genes. This development raises the prospect of important benefits, such as the ability to accelerate vaccine development. At the same time, it raises genuine concerns, and so we must consider carefully the implications of this research.

I therefore request that the Presidential Commission for the Study of Bioethical Issues undertake, as its first order of business, a study of the implications of this scientific milestone, as well as other advances that may lie ahead in this field of research. In its study, the Commission should consider the potential medical, environmental, security, and other benefits of this field of research, as well as any potential health, security or other risks. Further, the Commission should develop recommendations about any actions the Federal government should take to ensure that America reaps the benefits of this developing field of science while identifying appropriate ethical boundaries and minimizing identified risks. My Science and Technology Advisor, Dr. John P. Holdren, will be in communication with you about the scope and progress of your study.

1 ask that the Commission complete its study within six months and provide me with a report with its findings, as well as any recommendations and suggestions for future study that the Commission deems appropriate. Given the importance of this issue, I request that the Commission consult with a range of constituencies, including scientific and medical communities, faith communities, and business and nonprofit organizations.

It is vital that we as a society consider, in a thoughtful manner, the significance of this kind of scientific development. With the Commission's collective expertise in the areas of science, policy, and ethical and religious values, 1 am confident that it will carry out this responsibility with the care and attention it deserves.

Sincerely,

EXECUTIVE SUMMARY

The 21st century is widely heralded as the century of biology. Building on the fundamental understanding achieved in the second half of the last century, revolutionary advances are expected to improve many aspects of our lives, from clean energy and targeted, safer medicines to new industries. Prominent among emerging technologies is "synthetic biology," which aims to apply standardized engineering techniques to biology and thereby create organisms or biological systems with novel or specialized functions to address countless needs.

The idea of managing or manipulating biology to identify or develop specific characteristics is not new. Scientists have used DNA to create genetically engineered cells and organisms for many years; the entire biotechnology industry has grown around our expanding abilities in this area. The shelves of grocery stores across the United States are stocked with genetically engineered foods. Medical testing for genetically linked diseases is widely used by people across society.

By contrast, the idea of assembling living organisms wholesale from nonliving parts has intrigued human imagination for centuries with no success outside of fiction. For some, that possibility came one step closer last May with the announcement that scientists at the J. Craig Venter Institute had created the world's first self-replicating synthetic (human-made from chemical parts) genome in a bacterial cell of a different species. Intense media coverage followed, and the announcement ricocheted across the globe within hours as proponents and critics made striking claims about potential risks and benefits of this discovery and whether it amounted to an early-stage example of "creating life."

In response, President Barack Obama asked the Presidential Commission for the Study of Bioethical Issues (the Commission) to review the developing field of synthetic biology and identify appropriate ethical boundaries to maximize public benefits and minimize risks. The Commission approached this task through inclusive and deliberative engagement with a wide variety of sources, including scientists, engineers, faith-based and secular ethicists, and others who voiced, as expected, sometimes conflicting views on the science, ethics, and social issues surrounding synthetic biology. Through public meetings in Washington, D.C., Philadelphia, and Atlanta, the Commission created a forum for open dialogue to hear and assess competing claims about the science, ethics, and public policy relating to synthetic biology.

What the Commission found is that the Venter Institute's research and synthetic biology are in the early stages of a new direction in a long continuum of research in biology and genetics. The announcement last May, although extraordinary in many ways, does not amount to creating life as either a scientific or a moral matter. The scientific evidence before the Commission showed that the research relied on an existing natural host. The technical feat of synthesizing a genome from its chemical parts so that it becomes self-replicating when inserted into a bacterial cell of another species, while a significant accomplishment, does not represent the creation of life from inorganic chemicals alone. It is an indisputable fact that the human-made genome was inserted into an already living cell. The genome that was synthesized was also a variant of the genome of an already existing species. The feat therefore does not constitute the creation of life, the likelihood of which still remains remote for the foreseeable future. What remains realistic is the expectation that over time research in synthetic biology may lead to new products for clean energy, pollution control, and more affordable agricultural products, vaccines, and other medicines. The Commission therefore focused on the measures needed to assure the public that these efforts proceed with appropriate attention to social, environmental, and ethical risks.

President Obama gave the Commission a rare and exceptional opportunity in the world of presidential bioethics commissions to be forward looking instead of reactive. We are ahead of the emerging science, and this unique opportunity underscores the need for the government to act now to ensure a regular, ongoing process of review as the science develops. The Commission calls on the government to make its efforts transparent, to monitor risks, to support (through a peer-review process) the most publicly beneficial research, and to educate and engage with the public as this field progresses. The government must regularly review risk assessment and other issues as the science of synthetic biology progresses. Only through openness and active engagement with all the relevant communities will the government ensure ongoing public support and appropriate oversight. The Commission emphasizes the need to engage the public over time through improved science education, a publicly accessible fact-checking mechanism for prominent advances in biotechnology, and other efforts promoting clearer communication on the state of science.

Basic Ethical Principles for Assessing Emerging Technologies

To reach its recommendations, the Commission identified five ethical principles relevant to considering the social implications of emerging technologies: (1) public beneficence, (2) responsible stewardship, (3) intellectual freedom and responsibility, (4) democratic deliberation, and (5) justice and fairness. The principles are intended to illuminate and guide public policy choices to ensure that new technologies, including synthetic biology, can be developed in an ethically responsible manner.

The ideal of *public beneficence* is to act to maximize public benefits and minimize public harm. This principle encompasses the duty of a society and its government to promote individual activities and institutional practices, including scientific and biomedical research, that have great potential to improve the public's well-being. Public beneficence requires that when seeking the benefits of synthetic biology, the public and its representatives be vigilant about risks and harms, standing ready to revise policies that pursue potential benefits with insufficient caution.

The principle of *responsible stewardship* reflects a shared obligation among members of the domestic and global communities to act in ways that demonstrate concern for those who are not in a position to represent themselves (e.g., children and future generations) and for the environment in which future generations will flourish or suffer. Responsible stewardship recognizes the importance of citizens and their representatives thinking and acting collectively for the betterment of all. Importantly, it calls for *prudent vigilance*, establishing processes for assessing likely benefits along with assessing safety and security risks both before and after projects are undertaken. A responsible process will continue to assess safety and security as technologies develop and diffuse into public and private sectors. It will also include mechanisms for limiting their use when necessary.

Democracies depend on *intellectual freedom* coupled with the *responsibility* of individuals and institutions to use their creative potential in morally accountable ways. Sustained and dedicated creative intellectual exploration begets much of our scientific and technological progress. While many emerging technologies raise "dual use" concerns—when new technologies intended for good may be used to cause harm—these risks alone are generally insufficient to justify limits on intellectual freedom. As a corollary to the principle of intellectual freedom and responsibility, the Commission endorses a principle of *regulatory parsimony*, recommending only as much oversight as is truly necessary to ensure justice, fairness, security, and safety while pursuing the public good. This is particularly important in emerging technologies, which by their very definition are still in formation and are not well suited for sharply specified limitations. While clear guidelines to protect biosecurity and biosafety are imperative, undue restriction may not only inhibit the distribution of new benefits, but it also may be counterproductive to security and safety by preventing researchers from developing effective safeguards.

The principle of *democratic deliberation* reflects an approach to collaborative decision making that embraces respectful debate of opposing views and active participation by citizens. It calls for individuals and their representatives to work toward agreement whenever possible and to maintain mutual respect when it is not. Public discussion and debate with open interchange among all stakeholders can promote the perceived legitimacy of outcomes, even if those outcomes are unlikely to satisfy all interested parties. An inclusive process of deliberation, informed by relevant facts and sensitive to ethical concerns, promotes an atmosphere for debate and decision making that looks for common ground wherever possible and seeks to cultivate mutual respect where irreconcilable differences remain. It encourages participants to adopt a societal perspective over individual interests.

The principle of *justice and fairness* relates to the distribution of benefits and burdens across society. Biotechnology and emerging technologies such as synthetic biology, for good or ill, affect all persons. Emerging technologies like synthetic biology will have global impacts. For this reason, every nation has a responsibility to champion fair and just systems to promote wide availability of information and fairly distribute the burdens and benefits of new technologies.

Recommendations

With these guiding principles in mind, the Commission considered the array of public policy issues surrounding the emerging science of synthetic biology and makes the following recommendations. The reasons behind each recommendation are provided in the body of the report, and all readers are urged to consider carefully this more comprehensive account. In the cases of recommendations 1, 3, 5, 9, 11, 12, and 17, the Commission recommends ongoing review by the government, in consultation with the relevant scientific, academic, international, and public communities, with initial action completed within 18 months and made public. Some of these actions could easily be completed sooner, and the government is encouraged to do so and make its progress public.

Promoting Public Beneficence

Under the principle of public beneficence, the Commission recommends that the government review and make public findings regarding the scope of its research funding, especially for risk assessment and ethical and social issues raised by synthetic biology. This will promote public engagement and ensure needed transparency regarding federal efforts in the field of synthetic biology.

Recommendation 1: Public Funding Review and Disclosure

Through a central body such as the Executive Office of the President, the federal government should undertake a coordinated evaluation of current public funding for synthetic biology activities, including funding for research on techniques for risk assessment and risk reduction, and for the study of ethical and social issues raised by synthetic biology. This review should be completed within 18 months and the results made public.

Most potential products of synthetic biology are in very early stages of development. Therefore, basic research is critical to further expansion of this science and its effective translation into useful products. Necessary funding decisions should be made with the goal of advancing the public good, whether these decisions support synthetic biology research or other fields. The Commission does not offer an opinion on the relative merits of particular research directions, but recommends that such decisions receive ongoing evaluation as to the state of the science and its potential applications.

Recommendation 2: Support for Promising Research

Advancing the public good should be the primary determinant of relative public investment in synthetic biology versus other scientific activities. The National Institutes of Health, the Department of Energy, and other federal agencies should continue to evaluate research proposals through peer-review mechanisms and other deliberative processes created to ensure that the most promising scientific research is conducted on behalf of the public.

Information sharing is a critical mechanism for promoting scientific progress and innovation. The principle of public beneficence requires researchers, inventors, patent holders, and others to work together to develop creative strategies to maximize opportunities for innovation. The government should consider best practices and other policy guidance, if needed, to ensure that access to basic research results and tasks is not unduly limited.

Recommendation 3: Innovation Through Sharing

Synthetic biology is at a very early stage of development, and innovation should be encouraged. The Executive Office of the President, as part of the coordinated approach urged in Recommendation 4, should lead an effort to determine whether current research licensing and sharing practices are sufficient to ensure that basic research results involving synthetic biology are available to promote innovation, and, if not, whether additional policies or best practices are needed. This review should be undertaken with input from the National Institutes of Health, other agencies funding synthetic biology research, such as the Department of Energy and the National Aeronautics and Space Administration, the U.S. Patent and Trademark Office, industry, academia, and public civil society groups. The review should be completed within 18 months and the results made public.

Promoting Responsible Stewardship

The Commission endorses neither a moratorium on synthetic biology until all risks are identified and mitigated, nor unfettered freedom for scientific exploration. Instead, the Commission believes that the field of synthetic biology can proceed responsibly by embracing a middle ground—an ongoing process of prudent vigilance that carefully monitors, identifies, and mitigates potential and realized harms over time. Responsible stewardship requires clarity, coordination, and accountability across the government. While new agencies, offices, or authorities are not necessary at this time, the Executive Office of the President should lead an interagency process to identify and clarify, if needed, existing oversight authorities and ensure that the government is informed on an ongoing basis about developments, risks, and opportunities as this field grows. This process must be undertaken by an office with sufficient authority to bring together all parts of the government with a stake in synthetic biology and be sufficiently authoritative to effectively engage or oversee engagement with foreign governments.

Recommendation 4: Coordinated Approach to Synthetic Biology

The Commission sees no need at this time to create additional agencies or oversight bodies focused specifically on synthetic biology. Rather, the Commission urges the Executive Office of the President, in consultation with relevant federal agencies, to develop a clear, defined, and coordinated approach to synthetic biology research and development across the government. A mechanism or body should be identified to: (1) leverage existing resources by providing ongoing and coordinated review of developments in synthetic biology, (2) ensure that regulatory requirements are consistent and non-contradictory, and (3) periodically and on a timely basis inform the public of its findings. Additional activities for this coordinating body or process are described in other recommendations.

Because synthetic biology poses some unusual potential risks, as "amateur" or "do-it-yourself" (DIY) scientists and others outside of traditional research environments explore the field, these risks must be identified and anticipated, as they are for other emerging technologies, with systems and policies to assess and respond to them while supporting work toward potential benefits.

Recommendation 5: Risk Assessment Review and Field Release Gap Analysis

Because of the difficulty of risk analysis in the face of uncertainty—particularly for low-probability, potentially high-impact events in an emerging field—ongoing assessments will be needed as the field progresses. Regulatory processes should be evaluated and updated, as needed, to ensure that regulators have adequate information. As part of the coordinated approach urged in Recommendation 4, the Executive Office of the President should convene an interagency process to discuss risk assessment activities, including reasons for differences and strategies for greater harmonization across the government. It should also identify any gaps in current risk assessment practices related to field release of synthetic organisms. These reviews should be completed within 18 months and the results made public.

Coordination and careful risk analysis are essential steps for responsible stewardship, but they are not sufficient. There are several additional approaches, which are known today and continue to evolve as our abilities in this field grow, to limit uncertain risks in synthetic biology. Technology can be harnessed to build in safeguards. A number of safety features can be incorporated into synthetic organisms to control their spread and life span. Surveillance or containment of synthetic organisms is a concrete way to embrace responsible stewardship.

Recommendation 6: Monitoring, Containment, and Control

At this early stage of development, the potential for harm through the inadvertent environmental release of organisms or other bioactive materials produced by synthetic biology requires safeguards and monitoring. As part of the coordinated approach urged in Recommendation 4, the Executive Office of the President should direct an ongoing review of the ability of synthetic organisms to multiply in the natural environment and identify, as needed, reliable containment and control mechanisms. For example, "suicide genes" or other types of self-destruction triggers could be considered in order to place a limit on their life spans. Alternatively, engineered organisms could be made to depend on nutritional components absent outside the laboratory, such as novel amino acids, and thereby controlled in the event of release.

The timing of deliberate release of synthesized organisms into the environment and the need to analyze risks prior to release raises special concern. We must proceed carefully, particularly when the probability or magnitude of risks are high or highly uncertain, because biological organisms may evolve or change after release. For any field release, there must be adequate consideration of risk.

Recommendation 7: Risk Assessment Prior to Field Release

Reasonable risk assessment should be carried out, under the National Environmental Policy Act or other applicable law, prior to field release of research organisms or commercial products involving synthetic biology technology. This assessment should include, as appropriate, plans for staging introduction or release from contained laboratory settings. Exceptions in limited cases could be considered, for example, in emergency circumstances or following a finding of substantial equivalence to approved products. The gap analysis described in Recommendation 5 should determine whether field release without any risk assessment is permissible and, if so, when.

Synthetic biology is an international enterprise. Oversight and regulatory mechanisms should adopt an analogous approach, so that the united States is involved in regular discussions with other national and transnational organizations so they may seek coordination and consistency when possible.

Recommendation 8: International Coordination and Dialogue

Recognizing that international coordination is essential for safety and security, the government should act to ensure ongoing dialogue about emerging technologies such as synthetic biology. As part of the coordinated approach urged in Recommendation 4, the Executive Office of the President, through the Department of State and other relevant agencies such as the Department of Health and Human Services and the Department of Homeland Security, should continue and expand efforts to collaborate with international governments, the World Health Organization, and other appropriate parties, including international bioethics organizations, to promote ongoing dialogue about emerging technologies such as synthetic biology as the field progresses.

Responsible conduct of synthetic biology research, like all areas of biological research, rests heavily on the behavior of individual scientists. Creating a culture of responsibility in the synthetic biology community could do more to promote responsible stewardship in synthetic biology than any other single strategy. There are actors in the world of synthetic biology, namely engineers, chemists, materials scientists, computer modelers, and others, who practice outside of conventional biological or medical research settings. These groups may not be familiar with the standards for ethics and responsible stewardship that are commonplace for those working in biomedical research. This poses a new challenge regarding the need to educate and inform synthetic biologists in all communities about their responsibilities and obligations, particularly with regard to biosafety and biosecurity.

Recommendation 9: Ethics Education

Because synthetic biology and related research cross traditional disciplinary boundaries, ethics education similar or superior to the training required today in the medical and clinical research communities should be developed and required for all researchers and student-investigators outside the medical setting, including in engineering and materials science. As part of the coordinated approach urged in Recommendation 4, the Executive Office of the President, in consultation with the National Academy of Sciences, the National Academy of Engineering, the scientific community, and the public, should convene a panel to consider appropriate and meaningful training requirements and models. This review should be completed within 18 months and the results made public.

Additionally flowing from the principle of responsible stewardship, the Commission observed that careful and deliberate attention should be paid to discussions of potential moral objections as the field advances. Such moral objections include concerns that synthetic biology may conflict with essential conceptions of human agency and life; that its overall impact may be harmful to biodiversity, ecosystems, or food and energy supplies; and that it may fail to respect the proper relationship between humans and nature. The Commission devoted particular time and attention to discussing these possible moral objections during its deliberations. It heard relatively few objections from religious or secular ethicists concerning the present status of the field. Although the field currently is capable of significant but limited technical achievements, potential developments might raise further moral objections—for example, applications relying on the synthesis of genomes for higher order or complex species. Current objections to synthetic biology on moral grounds are often based on concerns regarding activities that the field is currently incapable of carrying out. However, continued evaluation and efforts to reach and maintain consensus will be needed as this field develops.

Recommendation 10: Ongoing Evaluation of Objections

Discussions of moral objections to synthetic biology should be revisited periodically as research in the field advances in novel directions. Reassessment of concerns regarding the implications of synthetic biology for humans, other species, nature, and the environment should track the ongoing development of the field. An iterative, deliberative process, as described in Recommendation 14, allows for the careful consideration of moral objections to synthetic biology, particularly if fundamental changes occur in the capabilities of this science and its applications.

Promoting Intellectual Freedom and Responsibility

The principle of intellectual freedom and responsibility asserts that restrictions on research, whether by self-regulation by scientists or by government intervention, should limit the free pursuit of knowledge only when the perceived risk is too great to proceed without limit. A moratorium at this time on synthetic biology research would inappropriately limit intellectual freedom. Instead, the scientific community—in academia, government and the private sector—should continue to work together to evaluate and respond to known and potential risks of synthetic biology as this science evolves. This effort may require the government to expand current oversight or engagement activities with non-institutional researchers. National Institutes of Health or the Department of Energy, for example, could be charged to sponsor education programs and workshops that bring together these groups. They could fund training grants or related programs to promote a culture of responsibility among this community. To exercise the appropriate level of oversight, the government will need to monitor the growth and capacity of researchers outside of institutional settings.

Recommendation 11: Fostering Responsibility and Accountability

The government should support a continued culture of individual and corporate responsibility and self-regulation by the research community, including institutional monitoring, enhanced watchfulness, and application of the National Institutes of Health Guidelines for Recombinant DNA Research. As part of the coordinated approach urged in Recommendation 4, the Executive Office of the President should evaluate, and re-evaluate periodically, the effectiveness of current research oversight mechanisms and determine what, if any, additional steps should be taken to foster accountability at the institutional level without unduly limiting intellectual freedom. Academic and private institutions, the public, the National Institutes of Health, and other federal funders of synthetic biology research should be engaged in this process. An initial assessment should be completed within 18 months and the results made public.

The norms of safe and responsible conduct that have evolved over time for many researchers in institutional settings may not be understood or followed by those new to the field or outside of these settings. It is important to note that presently there appears to be no serious risk of completely novel organisms being constructed in non-institutional settings including in the DIY community. Scrutiny is required to ensure that DIY scientists have an adequate understanding of necessary constraints to protect public safety and security, but at present the Commission sees no need to impose unique limits on this group.

Recommendation 12: Periodic Assessment of Security and Safety Risks

Risks to security and safety can vary depending on the setting in which research occurs. Activities in institutional settings, may, though certainly do not always, pose lower risks than those in non-institutional settings. At this time, the risks posed by synthetic biology activities in both settings appear to be appropriately managed. As the field progresses, however, the government should continue to assess specific security and safety risks of synthetic biology research activities in both institutional and non-institutional settings including, but not limited to, the "do-it-yourself" community. As part of the coordinated approach urged in Recommendation 4, the Executive Office of the President, working with the Department of Homeland Security, the Federal Bureau of Investigation and others, should undertake and periodically update this assessment. An initial review should be completed within 18 months and the results made public to the extent permitted by law.

Certain risks—generally involving national security—often warrant additional protections. Completely free exchange of data and materials might endanger public safety, but unilateral action to limit exchange could damage American research efforts in synthetic biology if u.S. scientists and students are excluded from full collaboration with the international community. Several recent advisory groups have recommended ongoing discussions among research universities, industry, and government on this topic. The Commission agrees that scientists should be actively engaged in these debates.

Recommendation 13: Oversight Controls

If the reviews called for in Recommendation 12 identify significant unmanaged security or safety concerns, the government should consider making compliance with certain oversight or reporting measures mandatory for all researchers, including those in both institutional and non-institutional settings, regardless of funding sources. It may also consider revising the Department of Commerce's export controls. Any such change should be undertaken only after consultation with the scientific, academic, and research communities and relevant science and regulatory agencies such as the National Institutes of Health, the Department of Homeland Security, and the Environmental Protection Agency. Export controls should not unduly restrain the free exchange of information and materials among members of the international scientific community.

Promoting Democratic Deliberation

Through democratic deliberation, questions about synthetic biology can be explored and evaluated on an ongoing basis in a manner that welcomes the respectful exchange of opposing views. This principle yields several opportunities for government and non-government actors alike to work together to ensure that synthetic biology advances in ways that respect divergent views and that avoid some of the misunderstanding and confusion, which at times, have hampered other scientific endeavors. To enhance democratic deliberation and thereby ensure that the progress in synthetic biology is widely understood and policy choices are thoughtfully considered, the Commission makes the following recommendations.

Recommendation 14: Scientific, Religious, and Civic Engagement

Scientists, policy makers, and religious, secular, and civil society groups are encouraged to maintain an ongoing exchange regarding their views on synthetic biology and related emerging technologies, sharing their perspectives with the public and with policy makers. Scientists and policy makers in turn should respectfully take into account all perspectives relevant to synthetic biology.

Recommendation 15: Information Accuracy

When discussing synthetic biology, individuals and deliberative forums should strive to employ clear and accurate language. The use of sensationalist buzzwords and phrases such as "creating life" or "playing God" may initially increase attention to the underlying science and its implications for society, but ultimately such words impede ongoing understanding of both the scientific and ethical issues at the core of public debates on these topics. To further promote public education and discourse, a mechanism should be created, ideally overseen by a private organization, to fact-check the variety of claims relevant to advances in synthetic biology.

This publicly accessible fact-check mechanism is among the most concrete ways by which public perception and acceptance of emerging technologies could be improved. Education also plays a key role in building public support for otherwise unfamiliar technologies. In light of our Nation's dependence on socially responsible scientific innovation for economic progress and individual well-being, the urgency of expanding effective science and ethics education cannot be exaggerated. Dialogue among individuals and public, private, and community groups demonstrates that science and its oversight do not belong exclusively to experts, highly trained professionals, or government officials. Science is a shared resource, affecting and belonging to all citizens.

Recommendation 16: Public Education

Educational activities related to synthetic biology should be expanded and directed to diverse populations of students at all levels, civil society organizations, communities, and other groups. These activities are most effective when encouraged and supported by various sources, not only government, but also private foundations and grassroots scientific and civic organizations. As part of the coordinated approach urged in Recommendation 4, the Executive Office of the President, with input from the scientific community, the public, and relevant private organizations, should identify and widely disseminate strategies to promote overall scientific and ethical literacy, particularly as related to synthetic biology, among all age groups.

Promoting Justice and Fairness

The principle of justice and fairness, at this very early stage of synthetic biology, yields two general recommendations that can be applied to both this technology and other emerging technologies. It directs those in government to consider rules for distribution of risks and benefits in research, and it directs those both in and outside of government to consider processes for just distribution of benefits and risks.

Recommendation 17: Risks in Research

Risks in research should not be unfairly or unnecessarily borne by certain individuals, subgroups, or populations. As part of the coordinated approach urged in Recommendation 4, the Executive Office of the President should lead an interagency evaluation of current requirements and alternative models to identify mechanisms that ensure that the risks of research in synthetic biology, including for human subjects and other affected parties, are not unfairly or unnecessarily distributed. Relevant scientific, academic, and research communities, including those in the private sector, should be consulted. This review should be completed within 18 months and the results made public.

Recommendation 18: Risks and Benefits in Commercial Production and Distribution

Risks to communities and the environment should not be unfairly distributed. Manufacturers and others seeking to use synthetic biology for commercial activities should ensure that risks and potential benefits to communities and the environment are assessed and managed so that the most serious risks, including long-term impacts, are not unfairly or unnecessarily borne by certain individuals, subgroups, or populations. These efforts should also aim to ensure that the important advances that may result from this research reach those individuals and populations who could most benefit from them. As part of the coordinated approach urged in Recommendation 4, the Executive Office of the President should evaluate current statutory mandates or regulatory requirements for distribution of risks and benefits and consider developing guidance materials and voluntary recommendations to assist manufacturers as appropriate.

In summary, the ability to easily manufacture and manipulate DNA in the laboratory has enhanced scientists' productivity and opened new directions for scientific exploration. In the future, scientists may be able to create entirely new organisms and systems previously unknown in the world today. But breakthroughs such as this raise a host of complex and sometimes controversial issues. They can help humanity in many ways, but they invariably carry some risks and often raise public concerns and fears. With these unprecedented achievements comes an obligation to consider carefully both the promise and potential perils that they could realize.

The recommendations detailed in this report provide a publicly accountable basis for ensuring that the field of synthetic biology advances to improve human health and public welfare with processes in place to identify, assess, monitor, and mitigate risks on an ongoing basis as the field matures. Risk assessment should precede field release of the products of synthetic biology. Ongoing assessment and review is required in several areas to avoid unnecessary limits on science and social progress, and to ensure appropriate restrictions to protect individual safety and our shared environment. Ongoing dialogue about concerns regarding the implications of synthetic biology for humans, other species, nature, and the environment should continue as synthetic biology develops from its infancy to a fully mature field of scientific inquiry and innovation.

INTRODUCTION

On May 20, 2010, the J. Craig Venter Institute announced it had created the world's first self-replicating synthetic genome in a bacterial cell of a different species.[1] Although scientists have used recombinant DNA techniques to engineer pieces of the genetic code for many years, this achievement marked the first time that all of the natural genetic material in a bacterial cell was replaced with a synthetic (i.e., human made or chemically synthesized) copy of the genes necessary for that organism to function. This announcement made headlines around the globe. Reaction was immediate, and it spanned the spectrum from expressions of enthusiasm to cries of alarm. Thoughtful deliberation about the meaning of this achievement was impossible in the hours that elapsed between the breaking news and the initial round of commentaries that ensued.

There is general agreement that this first self-replicating synthetic genome is an exceptional achievement, but there is also vigorous debate about just how momentous the Venter Institute's success is. Some scientists consider it a quantum leap; others see it as an incremental stride.[2] Whether one considers the accomplishment a major advance, a more modest technical step, or some combination of the two, one cannot deny the importance of understanding the potential implications of this and related accomplishments for humankind. The ability to synthesize vaccines, drugs, biofuels, and crops could do much to advance human welfare. At the same time, these innovations raise concerns about what we do not know—that is, whether there are attendant human or environmental risks—and what we perhaps should not know, that is, how to engineer forms of "life" to serve our own purposes.

Rather than offer an immediate opinion on the possible ethical and public policy implications—both positive and negative—of this scientific and technical accomplishment, President Barack Obama asked the Presidential Commission for the Study of Bioethical Issues (the Commission) as its first order of business to recommend how the developing field of synthetic biology and related biotechnologies can best maximize public benefits, minimize risks, and observe appropriate ethical boundaries.[3] He turned to the Commission to conduct "a study of the implications of this scientific milestone, as well as other advances that may lie ahead in this field of research." It was directed to consider the "potential medical, environmental, security, and other benefits of this field of research, as well as any potential health, security, or other risks." The President charged the Commission to provide recommendations within six months on "any actions the federal government should take to ensure that America reaps the benefits of this developing field of science while identifying appropriate ethical boundaries and minimizing identified risks." Much stands to be gained by the government taking a deliberative and open approach to decision making in this and many other complex scientific and technical areas of public importance.

Recent advances in biotechnology have transformed the life sciences, yielding a level of innovation rarely witnessed in human history. These achievements raise a host of complex and often controversial issues. Breakthroughs can help humankind in many ways, but they invariably carry some risks. Discoveries of new ways of improving or enhancing life raise public hopes and expectations, but they also raise public concerns and, often, fears. Proponents of synthetic biology cite its potential to reduce our reliance on fossil fuels and transform medical care and human health, among other possible benefits. Critics express concerns about "playing God," threatening biodiversity and the organization and natural history of species, demeaning and disrespecting the meaning of life, and threatening longstanding concepts of nature. With these unprecedented opportunities and achievements comes an obligation to consider carefully both the promise and potential perils that they could realize.

Airing these expectations and concerns in a public forum maximizes the potential for public benefit and illuminates risks and possible harms— physical, environmental, and social—that deserve our attention and careful consideration. In addressing the President's charge the Commission therefore attempted to be an inclusive and deliberative body, encouraging the exchange of well-reasoned perspectives with the goal of making recommendations that will serve the public well and will advance the public good. It gathered specific information about the state of synthetic biology, reviewed the findings and recommendations of numerous u.S. and international groups, and listened to sometimes conflicting scientific, ethical, and social perspectives. It sought common ground where possible and generally found it. When common ground was impossible to find, the Commission cultivated mutual respect through active engagement with differing views.

The Commission's Process

In conducting its work, the Commission invited experts and representatives of the public to explore contested territory from multiple perspectives. Some guests presented information about recent and upcoming achievements in the science of synthetic biology, including current and future applications and benefits. Others shared their perspectives on anticipated risks, related regulatory and oversight issues, and ethical considerations. The Commission solicited questions from the public as well as from its own members. This format contributed to highly interactive and valuable sessions. In addition, the Commission encouraged the public to provide written comments throughout its deliberations, and nearly 40 individuals and groups submitted comments. It also consulted with relevant federal agencies and private entities considering similar questions.

Formal deliberations began with an overview of potential benefits. Without any realistic promise of benefits, no risks would be worth taking. Expert panelists cited a host of potential benefits including more efficient and effective drug development; accelerated synthesis of vaccines in response to pandemics; and the ability to engineer algae and other microbes to spur advances in clean-burning fuel, agriculture, bioremediation, and medicine. The Commission also heard about the promise of a robust bio-economy beginning to materialize in the form of novel technological platforms. These and other areas of research in synthetic biology offer significant opportunities for economic growth and job creation.

After discussing the possible benefits of synthetic biology, the Commission considered the current and foreseeable risks posed by this rapidly evolving fi eld. Although the risks at this early stage in the field's development are well managed and relatively small in comparison to the anticipated benefits of the fi eld, they do exist, and several themes emerged in Commission discussions.

First, sheer prudence suggests that we as a society must respect the intricacies of the natural world. Biological systems have developed over billions of years, and their interactions with the environment are astoundingly complex. We are far from being proficient speakers of the language of life, and our capacity to control synthetic organisms that we design and release into the world is promising but unproven.

Second, understanding our own limitations is an essential prelude to minimizing the risks that will accompany ongoing breakthroughs in synthetic biology and related fields. Like other new technologies, synthetic biology poses uncertain risks. Rapidity of change, both in the field of biology and in the public's understanding of it, as well as accelerating information exchange and technological competence heighten these concerns. Today, predicting cell function from gene sequence alone is very difficult and often impossible.[4] While the successful synthesis of a functional bacterial chromosome is an essential technological step for the development of synthetic biology, it represents a preliminary advance. We remain far from having the scientific and technical expertise required to create truly novel functioning organisms. We must be cognizant, however, of our limited current understanding of what synthetic biology and related technologies may produce in the future and be willing to reassess benefits and harms as the field develops.

Third, ancillary effects and challenges should be recognized and considered. The rise of an economy based on biotechnology may expand jobs and lead to significant financial benefits, but it could also result in economic displacement, excessive demands on already scarce resources, and increased social and economic stratification. Anticipating all of the ramifications of our actions is impossible, but determining how to respond to this uncertainty is the better part of wisdom. The Commission also considered related questions regarding how the u.S. government can best respect intellectual freedom in scientific inquiry and nurture the development of synthetic biology in a way that maximizes its potential benefits while reducing the risks and likelihood of direct and indirect harms.

Critical to all of these themes is the importance of earning public trust in the integrity of both the scientific and engineering communities and the applicable regulatory systems. The Commission therefore focused on the need for greater public education and engagement on these issues as a prerequisite for public acceptance of this new technology and assurance of constructive criticism moving forward.

Basic Ethical Principles for Assessing Emerging Technologies

In approaching its task, the Commission was mindful of the need for an ethical framework for considering the implications of new and emerging technologies like synthetic biology, which itself represents one step in a long continuum of scientific innovation.[5] This is a unique opportunity to consider the ethics of an emerging technology at a very early stage of its development.[6] The Commission found many efforts to shape policy, governance, and regulation related to synthetic biology, but few examples of an ethical framework upon which to gird such proposals. Accordingly, in weighing alternative policy preferences and perspectives, it identified five ethical principles relevant to considering the social implications of synthetic biology as well as all emerging technologies. These principles provide a useful vehicle through which to evaluate the current state of the field and formulate the Commission's recommendations.

The guiding principles are: (1) public beneficence, (2) responsible stewardship, (3) intellectual freedom and responsibility, (4) democratic deliberation, and (5) justice and fairness. These principles should be understood as provisional guideposts. The Commission encourages others to subject these principles, and the recommendations based on them, to further refinements and revisions, as it has done and will continue to do in the future.

Public Beneficence

The ideal of public beneficence is to act to maximize public benefits and minimize public harm. The principle encompasses the duty of a society and its government to promote individual activities and institutional practices, including scientific and biomedical research, that have great potential to improve the public's well-being. In the case of emerging technologies like synthetic biology, this improvement may be by means of providing improved or more widely available forms of medical and health care, food, shelter, transportation, clothing, and eco-friendly fuel, along with other means of improving people's lives. Scientific and technological discovery often have the added potential of increasing economic opportunities, which also redound to the public good.

The *Belmont Report*, a landmark statement of ethical principles for research involving human subjects, defined beneficence to require that "[p]ersons are treated in an ethical manner not only by respecting their decisions and protecting them from harm, but also by making efforts to secure their wellbeing."[7] Two general rules stem from this principle: first, do no harm; and second, maximize possible benefits and minimize possible harms.

For synthetic biology and other emerging technologies, we need to apply the principle of beneficence beyond the individual level, the primary emphasis of the *Belmont Report*, to the institutional, community, and public levels, while not overlooking possible harms and benefits to individuals. Policy makers should adopt a societal perspective when deciding whether to pursue particular benefits of synthetic biology research in the face of risks and uncertainty. When deciding whether to restrict these pursuits, a similar consideration of community interests and potential positive and negative impacts is essential.

Public beneficence requires that when seeking the benefits of synthetic biology, the public and its representatives be vigilant about risks and harms, standing ready to revise policies that pursue potential benefits with insufficient caution. The Commission explores the concomitant challenges of meaningful and valuable risk-benefit analysis and potential strategies to address them in the "Responsible Stewardship" section, below.

Responsible Stewardship

Among living beings, humans are in a unique position to be responsible stewards of nature, the earth's bounty, and the world's safety. Human society and governments have a duty to proceed prudently in promoting science and technologies, many of which can improve human welfare but also can harm the environment, create security risks, or otherwise lead to adverse consequences for vulnerable populations or future generations. The principle of responsible stewardship reflects a shared obligation among members of the domestic and global communities to act in ways that demonstrate concern for those who are not in a position to represent themselves (e.g., children and future generations) and for the environment in which future generations will flourish or suffer.

Scientists, policy makers, and the public are tasked with appreciating that the tools of science and technology possess both remarkable potential to enhance future lives and a spectrum of risks capable of causing harm. Both demand attention and action.

Responsible stewardship recognizes the importance of citizens and their representatives thinking and acting collectively for the betterment of all, especially those who cannot represent themselves. These activities must respect the significant impact—both positive and negative—that our decisions have on our world, both today and in the future.

Benefits and risks extend to humans, nonhuman species, and the environment, each with unique needs and vulnerabilities. Emerging technologies present particularly profound challenges for responsible stewardship because our understanding of these potential benefits and risks is largely incomplete, preliminary, and uncertain. The prospect of intentional misuse by malicious actors further complicates efforts to respond adequately to the spectrum of benefits and risks.

Responsible stewardship addresses these varied challenges by calling for actions that embrace potential benefits while mitigating risks over time and across all populations. It calls for broader risk-benefit discussions than what would typically be required based on a concern for public beneficence alone. The principle of responsible stewardship rejects two extreme approaches: an extreme action-oriented approach that pursues technological progress without limits or due regard for public or environmental safety, and an extreme precautionary approach that blocks technological progress until all possible risks are known and neutralized. While the action-oriented approach is irresponsibly brazen, the precautionary approach is overly wary. Both fail to carefully assess the most likely and significant benefits against the most likely and significant harms. Through the development of agile, measured oversight mechanisms, responsible stewardship rejects positions that forsake potential benefits in deference to absolute caution and those that ignore reasonably foreseeable risks to allow unfettered scientific exploration.

This principle is applied to synthetic biology and other emerging technologies through open decision-making processes informed by the best available science. Responsible stewardship calls for *prudent vigilance*, establishing processes for assessing likely benefits along with safety and security risks both before and after projects are undertaken. A responsible process will continue to evaluate safety and security as technologies develop and diffuse into public and private sectors. It will also include mechanisms for limiting their use when indicated.

Prudent vigilance does not demand extreme aversion to all risks. Not all safety and security questions can be definitively answered before projects begin, but prudent vigilance does call for ongoing evaluation of risks along with benefits. The iterative nature of this review is a key feature of responsible stewardship. It recognizes that future developments demand that decisions be revisited and amended as warranted by additional information about risks and potential benefits. The duty to be responsible stewards of nature, the earth's bounty, and the world's safety rests on concern not only for human health and well-being today but also, and importantly, for future generations and the environment looking forward.

Intellectual Freedom and Responsibility

Democracies depend on intellectual freedom coupled with the responsibility of individuals and institutions to use their creative potential in morally responsible ways. Sustained and dedicated creative intellectual exploration begets much of our scientific and technological progress. Without the free marketplace of ideas we would not have many of the scientific discoveries and advancements that have aided us in harnessing energy, sustaining life, and raising our collective standard of living. Intellectual freedom, therefore, is critical for developing innovative technologies that can compete in the global marketplace, and it is a necessary condition for industrial and academic collaborations that yield useful products and tools. While many emerging technologies raise concerns about their potential malevolent use, these risks alone are generally insufficient to justify limits on intellectual freedom. If we as a society stifle intellectual freedom for fear of enabling harm, we will be unprepared and vulnerable if that harm is unleashed upon us. A robust public policy regarding the responsible conduct of science must promote the creative spirit of scientists and unambiguously protect their intellectual freedom.

At the same time, responsible science should reject the technological imperative: the mere fact that something new can be done does not mean that it ought to be done. The history of science here and abroad is sadly full of examples of intellectual freedom exercised without responsibility that resulted in appalling affronts to vulnerable populations, the environment, and the ideals of the profession of science itself. Scientists who act irresponsibly are capable not only of harming themselves and other individuals, but also of harming their communities, their nations, and international relations. Society as a whole has a stake in what scientists and engineers do, and they must not operate as if their research is totally independent of the groups who will experience both the benefits and burdens of their work. Risks may be especially great when those who provide the means and those who experience benefits are not the same. It is society that collectively provides the means for scientists to do their work and it is to society collectively that scientists bear profound responsibility.

As a corollary to the principle of intellectual freedom and responsibility, the Commission endorses a principle of regulatory parsimony, recommending only as much oversight as is truly necessary to ensure justice, fairness, security, and safety while pursuing the public good. Regulatory parsimony is particularly important in emerging technologies, which by their very definition are still in formation and are not always well-suited for sharply specified limitations. The blunt instruments of statutory and regulatory restraint may not only inhibit the distribution of new benefits, but they can be counterproductive to security and safety by preventing researchers from developing effective safeguards.[8] With sufficient freedom to operate, tomorrow's achievements may render moot the risks of today. Self-regulation also promotes a moral sense of ownership within a professional culture of responsibility.

Democratic Deliberation

The principle of democratic deliberation reflects an approach to collaborative decision making that embraces respectful debate of opposing views and active participation by citizens. It calls for individuals and their representatives to work toward agreement whenever possible and to maintain mutual respect when it is not.[9]

At the core of democratic deliberation is an ongoing, public exchange of ideas, particularly regarding the many topics—in science and elsewhere—in which competing views are advocated, often passionately. Through formal and informal deliberative processes, decision makers and the people they represent should strive for mutually acceptable reasons to justify the policies that they adopt. These justifications should be expressed in ways that are accessible to those to whom such policies apply.

Citizens, individually and collectively, are active participants in democratic deliberation, engaging in dialogues both among themselves and with their representatives charged with developing policy. Public discussion and debate promote the legitimacy of whatever outcomes are reached, even if those outcomes are unlikely to please all interested parties. A process of active deliberation and justification promotes an atmosphere for debate and decision making that looks for common ground wherever possible and seeks to cultivate mutual respect where irreconcilable differences remain. It encourages participants to adopt a societal perspective over individual interests.

Importantly, democratic deliberation recognizes that while decisions must eventually be reached, those decisions need not (and often should not) be permanent, particularly when subsequent developments warrant additional examination. Democratic deliberation can correct the inevitable mistakes that arise when decisions are made collectively, provided that it is an ongoing, dynamic process. It recognizes the importance of challenging previously reached conclusions in light of new information or perspectives. It therefore requires citizens to take seriously the possibility that the views of one's opponents may be shown to be correct in the future and to be open to changing their own views.

With careful attention to the processes through which decisions are reached and justified, democratic deliberation promotes outcomes that are inclusive, thoughtfully considered, and respectful of competing views.

The principle of democratic deliberation, although a less familiar principle in bioethics than the principles of beneficence and justice, is particularly well-suited to the assessment of emerging technologies, including synthetic biology.[10] These fields offer the promise of remarkable potential benefits to science and society, yet they also raise risks regarding unintended consequences or possible malicious use. Each of these areas is clouded by uncertainty, complicating efforts to promote innovation while minimizing the likelihood of harm. Finding this balance demands careful ongoing review of the science and its applications. It presents an ideal opportunity for broad engagement and dialogue among the scientific community, policy makers, and the public. This active public engagement can enhance the decisions that are reached and the overall public understanding of them, as well as the related issues in science and technology that are central to the future of this new technology, as well as to our Nation and the world.

Justice and Fairness

The principle of justice and fairness relates to the distribution of benefits and burdens across society. Emerging technologies like synthetic biology, for good or ill, affect all persons. Society as a whole has a claim toward reasonable efforts on the part of both individuals and institutions to avoid unjust distributions of the benefits, burdens, and risks that such technologies bring. This same claim extends internationally to all those who may be affected—positively or negatively—by synthetic biology and its applications. As much as possible, and consistent with establishing essential incentives for creating new knowledge and translating it into vibrant markets, a fundamental principle of fairness suggests that society should seek to ensure that the benefits and burdens of new technologies are shared.

A commitment to justice and fairness is a commitment to seek to ensure that individuals and groups receive that to which they are entitled, that is, what they can reasonably and legitimately expect. Identifying, anticipating, and assessing what is reasonable to expect and determining how to measure and compare potential risks and benefits are complex activities, even in the best of circumstances and with the most complete data. They are made more difficult by the uncertainties surrounding scientific advances and the emergence of new technologies. How, for example, are we to measure and compare the benefits of a technological innovation that leads to an effective medical treatment available on an unprecedented scale at low cost against the costs imposed by the disruption and displacement of previously existing technologies and the people whose livelihoods depends upon them?

Advances produced through biotechnology can be highly beneficial but costly. How can and should we ensure that such advances reach those who could benefit most rather than being available only to those who can afford to pay? While such questions are difficult to answer, society must work to provide answers that are both just and fair.

The principle of justice and fairness also suggests that society should seek to ensure that the unavoidable burdens of technological advances do not fall disproportionately on any particular individual or group. Technological innovation benefits from public investment and from societal contribution toward safe and supportive research environments, and so it is reasonable that society expect a return on that investment.

Justice and fairness extend not only from individual societies to their constituents but also from individual societies to the international community overall. Emerging technologies like synthetic biology can and likely will have global impacts.

For that reason, every nation has a responsibility to champion fair and just systems to promote the widest availability of information, the broadest distribution of beneficial technologies, and the most expansive culture of responsibility for biosafety and biosecurity.

About This Report

With these guiding principles in mind, the Commission considered the array of ethical public policy issues surrounding the field of synthetic biology. It reviews the science and potential benefits of this field in Chapters 2 and 3. Chapter 4 summarizes the existing oversight framework for new and emerging technologies like synthetic biology. Chapter 5 examines the implications of synthetic biology as viewed through the five principles described above and offers recommendations to ensure that society reaps the benefits of this developing field of science while identifying appropriate ethical boundaries and minimizing identified risks.

SCIENCE OF SYNTHETIC BIOLOGY

Synthetic biology is the name given to an emerging field of research that combines elements of biology, engineering, genetics, chemistry, and computer science. The diverse but related endeavors that fall under its umbrella rely on chemically synthesized DNA, along with standardized and automat-able processes, to create new biochemical systems or organisms with novel or enhanced characteristics. Whereas standard biology treats the structure and chemistry of living things as natural phenomena to be understood and explained, *synthetic* biology treats biochemical processes, molecules, and structures as raw materials and tools to be used in novel and potentially useful ways, often quite independent of their natural roles. It joins the knowledge and techniques of biology with the practical principles and techniques of engineering. "Bottom-up" synthetic biologists, those in the very earliest stages of research, seek to create novel biochemical systems and organisms from scratch, using nothing but chemical reagents. "Top-down" synthetic biologists, who have been working for several decades, treat existing organisms, genes, enzymes, and other biological materials as parts or tools to be reconfigured for purposes chosen by the investigator.

For the purposes of this report, the Commission focused on the molecular and cellular engineering techniques of synthetic biology and the most foreseeable benefits of this very early field. In time, synthetic biology products for clean energy, pollution control, agriculture, and medicine, may change our lives and our shared environment through the development of novel applications. Because the potential applications of synthetic biology are speculative at this time, and the field is advancing in exciting directions, it is inviting both optimism and unease among scientists and the public.

From Molecular Biology to Synthetic Biology

Synthetic biology is deeply rooted in molecular biology, a field that emerged decades ago with the discovery of the structure and composition of DNA. DNA molecules provide the instructions that direct cell growth, development, and differentiation in every living organism. They contain a sequence of four types of chemical building blocks—adenosine, thymine, cytosine, and guanine (A, T, C, and G)—that combine, ladder-like and in various order, into "base pairs" that are combined into sets called "genes" (see Figure 1).

Individual gene sequences code for particular proteins, which are what enable cells to function. Collectively, the complete DNA sequence of an organism is called its "genome." Genome variation is what makes individual organisms unique.

Though not described as such at the time, the earliest achievements in what is today called synthetic biology can be traced to the birth of genetic engineering in the 1970s. Genetic engineering, sometimes called gene-splicing or recombinant DNA research, is the intentional manipulation of an organism's genetic material using tools that cut, move, and reattach (recombine) DNA segments within and across different organisms.

In 1972, Stanford University biochemist Dr. Paul Berg created the first recombinant DNA molecules by splicing DNA from a bacterial virus into that of a monkey virus, SV40.[1] Two years later, scientists created the first trans-genic mammal by introducing foreign DNA into mouse embryos.[2] Today, transgenic mice are a staple of biomedical research. They are used to regulate the expression of individual genes in order to understand how those genes interact with the environment and, in turn, affect human health. Using trans-genic mice also enables researchers to increase or decrease specific proteins and better understand their individual roles and functions.[3]

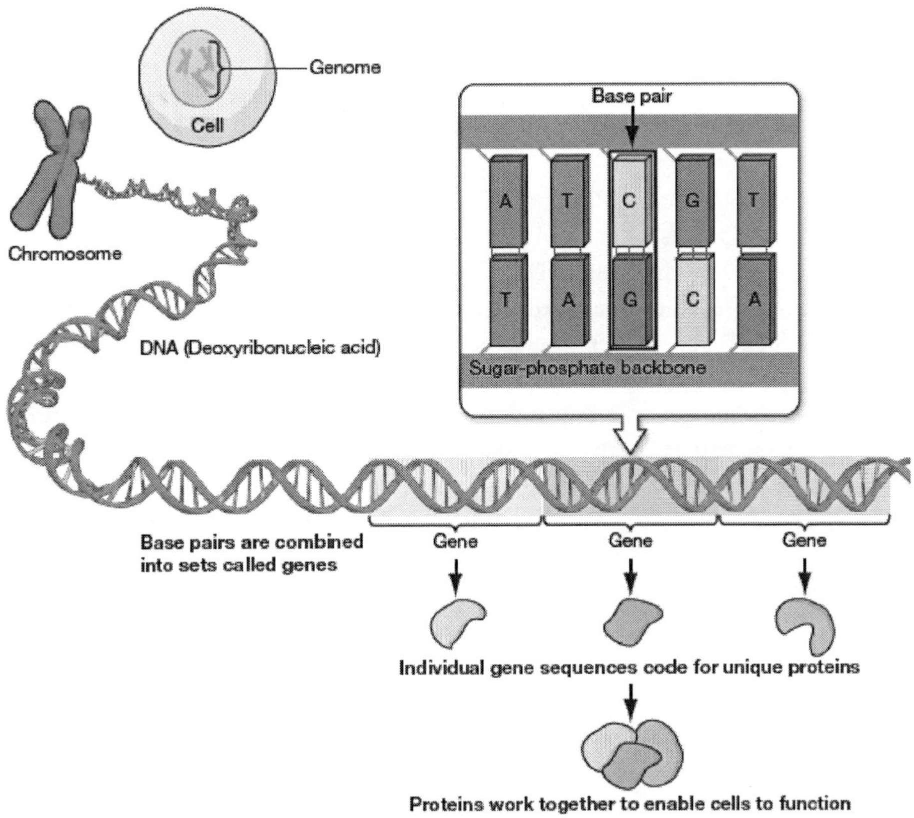

Figure 1. DNA, genes, and proteins.

As recombinant DNA technology began to develop in the 1970s, individual scientists, policy makers, and nations undertook profound debate about the safety and permissibility of this research—whether it was too dangerous to proceed at all—in the face of deep uncertainty.[4] Like synthetic biology today, great promise and potential risks were identified.[5] Expert and lay groups intensely debated concerns about possible adverse human health and environmental effects.

In 1974, a group of American scientists called for a moratorium on DNA research and the scientific community voluntarily obliged. To resolve this stalemate, in 1975 scientists from around the world, policy makers, lawyers, and press met together at the Asilomar Conference Center in Pacific Grove, California, to debate safety issues. The deliberations at the Asilomar Conference on Recombinant DNA led to formation of guidelines to ensure safety and a scientific peer review group, today known as the Recombinant DNA Advisory Committee of the National Institutes of Health (NIH). Both the guidelines and the Recombinant DNA Advisory Committee remain as critical components of the genetic engineering research oversight system (see Chapter 4 for further discussion). Many of the processes first proposed at the Asilomar Conference remain in place, though some have changed in the intervening years as understanding of risks has improved. Scientists and policy makers have pointed to Asilomar as valuable precedent when considering debates regarding research in synthetic biology.

By the end of the 1970s, scientists had created the first commercial product of genetic engineering. An extraordinary benefit for human health, human insulin produced using recombinant DNA technology transformed treatment for diabetes.[6] Following its entrance to the market, public acceptance of this new technology grew and fears decreased significantly.[7]

In the early 1980s, researchers developed another revolutionary technique, called polymerase chain reaction (PCR). The PCR method enabled researchers to amplify and make simple changes to DNA pieces. PCR acts like a molecular copy machine, allowing scientists to enlarge individual DNA sections and manipulate them more easily.

By the early 1990s, automated DNA sequencing became available. This technology considerably accelerated the process of determining the order of individual gene segments, called "nucleotides," or, when very small (typically less than 20 base pairs), "oligonucleotides."

Through large-scale genome sequencing efforts, primarily the public and private Human Genome Project, scientists were able identify the complete genetic codes of numerous naturally occurring organisms, including bacteria, viruses, and higher organisms such as mice and humans. The genome of a bacterial cell typically includes 5 to 10 million base pairs, although the synthesized genome of the bacteria in the J. Craig Venter Institute research, described below, contained just over 1 million base pairs.[8] By comparison, a fruit fly genome includes 165 million base pairs, and the human genome includes more than 3 billion base pairs.

These significant differences in scale help place the achievement of the Venter Institute team in context. While it represents the first successful synthesis of a complete genome of a single-celled bacterium, it is a relatively small genome compared to those of other species.

After scientists could sequence naturally occurring DNA, they developed techniques to synthesize, or chemically construct, DNA and pieces of DNA.[9]

Figure 2 shows an early DNA synthesis machine and the individual chemicals, including nucleic acids, used to construct sequences. Within the last few years, researchers have developed methods to accurately synthesize increasingly longer segments of DNA and to bring them together into even larger segments of DNA.

Stemming from this research, a small industry of commercial DNA synthesis providers has emerged. Five of the main companies, roughly 80 percent of the market, are based in the united States.[10]

The development of DNA synthesis technology has enabled scientists to make entire genes, and, eventually, the complete genome of a microorganism using synthetic methods alone. By synthesizing a complete genome for a bacterial cell and transferring it to a cell with its own genome that was later lost, researchers at the Venter Institute created a self-replicating bacterial cell with entirely chemically constructed DNA (see Figure 3).[11]

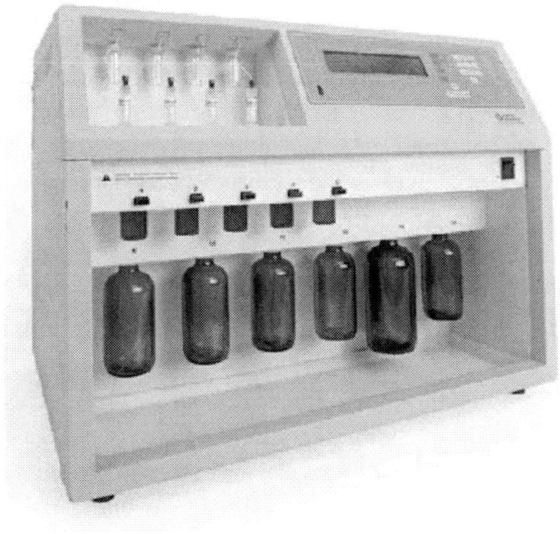

Figure 2. Early DNA synthesis machine. (Courtesy of Life Technologies).

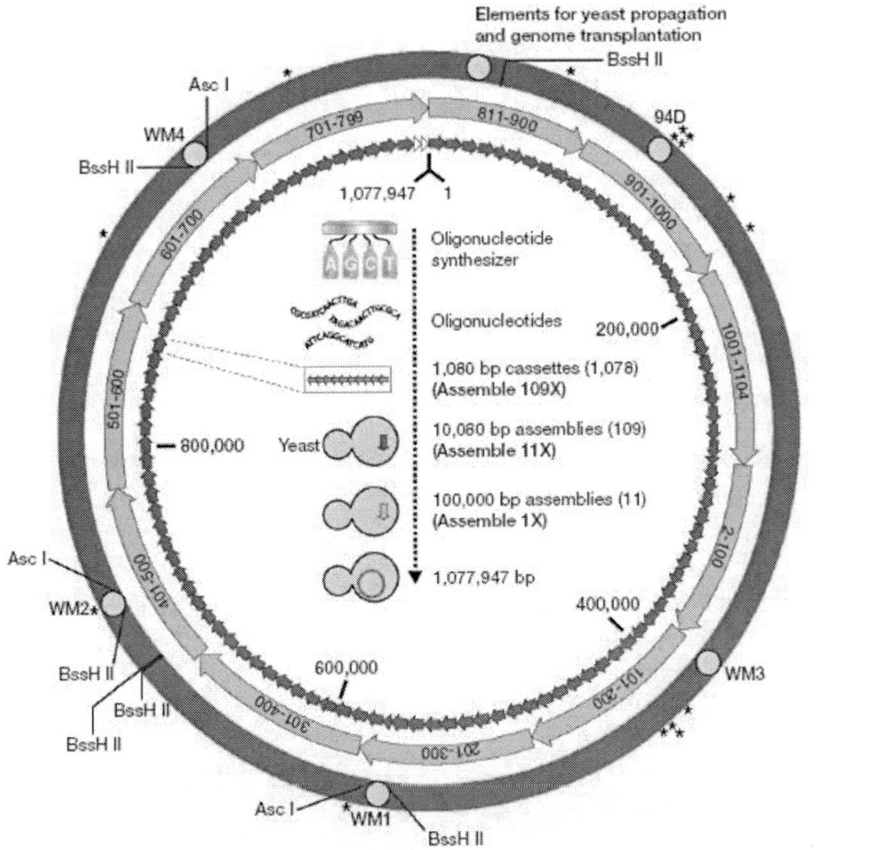

Figure 3. The assembly of a synthetic *M. mycoides* genome in yeast. Source: Gibson, D.G., et al. (2010). Creation of a bacterial cell controlled by a chemically synthesized genome. *Science* 329(5987):52-56.

Of note, many scientists observe that this achievement is not tantamount to "creating life" in a scientific sense because the research required a functioning, naturally occurring host cell to accept the synthesized genome. At the same time, this development should not be undersold. For many, this work represents the "proof of principle" that synthetic biology techniques can be used to construct cells and other organisms with novel characteristics.[12] While this small step does not give us the ability to grow larger-scale organisms, human tissue, or other tools of regenerative medicine, it is an incremental step on which future technical and scientific achievements will build.

The First Self-Replicating Synthetic Bacterial Cell

A May 21, 2010 publication in the journal *Science* by researchers from the Venter Institute announced the design, synthesis, and assembly of the 1.08 million base pair chromosome of a modified *Mycoplasma mycoides* bacterial genome. Beginning with an accurate, digitized genome of the bacteria, the researchers added four watermark sequences to identify the genome more clearly. They then designed more than 1,000 cassettes of DNA including approximately 1,080 base pairs each, with 80 base pair overlaps on each cassette representing adjacent sequences. The fragments were assembled sequentially in yeast. First, 10 cassettes each combined to make 10,000 base pair intermediates. Ten of those intermediates next were assembled to produce eleven 100,000 base pair intermediates, which were then combined into the complete genome. The newly synthesized genome was initially grown in yeast before being isolated and transplanted into cells of another bacterium, *Mycoplasma capricolum*. The genome of the recipient cells were lost as the cells were incubated, resulting in viable, self-replicating *Mycoplasma mycoides* cells containing only DNA from the synthetic genome.

Early molecular biology laid the groundwork for today's synthetic biology, but more recent technological advances have accelerated its development. First, scientists have developed the ability to mechanically synthesize increasingly longer DNA segments accurately and more rapidly than had been possible previously. Second, the costs for DNA synthesis have fallen dramatically over the past decade, dropping from about $30 to well under $1 per base pair.[13]

Computer modeling, not readily available until recently, is also facilitating the design of novel genetically engineered biological systems. As with electrical or civil engineering, modeling is intended to help scientists predict the behavior of a system before it is actually built. Although biological systems are not nearly as easily modeled as an electronic circuit or a bridge, at least at this time, sophisticated simulations, mostly in single-cell systems, are contributing to improved computer modeling of synthetic biological systems.

Synthetic Biology Techniques and Strategies

As discussed previously, to date synthetic biology has been characterized by top-down and bottom-up approaches.[14] The techniques overlap to some extent, and both approaches share a common goal: to engineer specific biological functions with predictability and reliability. In the future, these approaches may come together. For now, it is useful to consider both as illustrative of different experimental methods to reach the same goal.

Top-Down Approach

Through the top-down approach, in use since the 1970s, scientists use synthetic biology to re-design existing organisms or gene sequences with the goal of stripping out unnecessary parts, or replacing or adding specific parts to achieve new or amplified characteristics and functions (see Figure 4). Using this approach, scientists aim to remove parts of an organism or genetic code to create what some have dubbed a "chassis organism" that can then be modified through the addition or subtraction of engineered genetic circuits or metabolic pathways.[15]

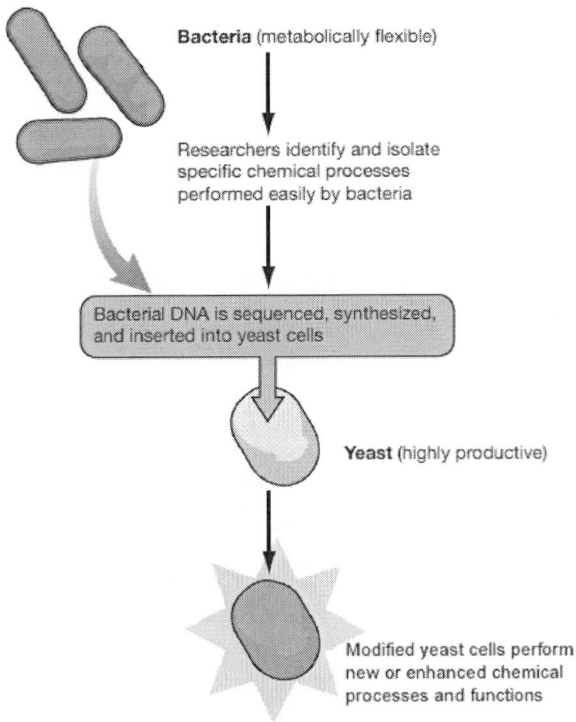

Figure 4. Example of a top-down approach to synthetic biology.

One recent example of the top-down approach in synthetic biology is the identification of a "minimal genome."[16] This research provided proof of principle that the total genetic material of a small bacterium, its genome, could be pared down into a functioning unit consisting of only a subset of the organism's original genes.

Top-down synthetic biology is also defined by borrowing properties from one or more living systems to create something new. One example is combining the productivity of yeast cells with the metabolic flexibility of bacteria. In this approach, researchers identify a range of chemical processes performed easily by various types of bacteria and insert these processing abilities into industry-standard yeast cells. In one case, the result was an efficient way to manufacture simple, yet high-value chemicals called methyl halides, used as agricultural fumigants and as fuel ingredients, starting with readily available plant matter such as corn stalks, sugar cane, and switchgrass.[17] Top-down synthetic biology is made easier through the use of increasingly accessible and inexpensive DNA sequencing and DNA synthesis technologies. Scientists can use them to "trawl" for bacterial genes that perform useful tasks and then copy and paste that DNA into yeast, without ever touching (or laboriously culturing) the bacteria, as was once required.[18]

Bottom-up Approach

In bottom-up synthetic biology, which is relatively new and significantly more challenging, scientists aim to build living systems from raw materials starting with non-living components. For example, a team of scientists is aiming to create completely artificial systems using only non-living materials that mimic the behavior of actual cells. The products of this research are called chemical cells, or "Chells."[19] Bottom-up approaches also include efforts to create genetically engineered circuits and switches to turn specific functions on and off in response to identified stimuli. In some cases, the bottom-up approach could theoretically result in an entirely new organism or material with functions that may be different from currently existing organisms or cells. In other cases, parts with known functions may function differently when assembled into a new material or organism.

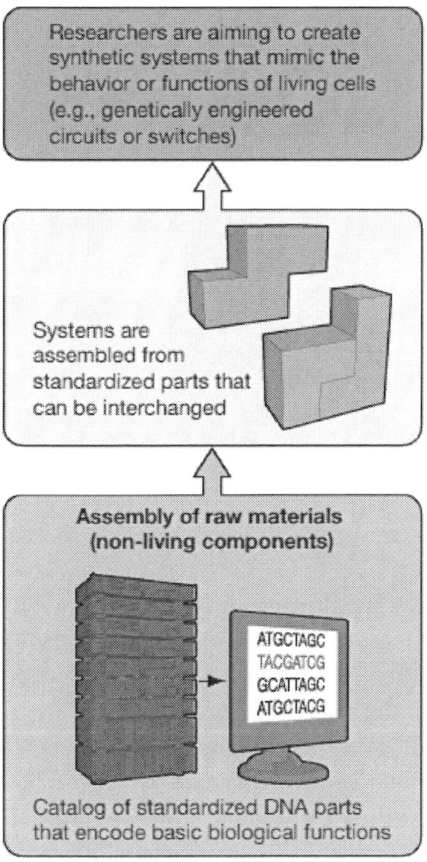

Figure 5. Example of a bottom-up approach to synthetic biology.

Bottom-up approaches are sometimes characterized by their reliance on assembling systems from chemically synthesized standardized parts that perform desired functions in a predictable manner and can be interchanged.[20] Like Legos® or computer components, a goal of this work is to develop a set of basic chemically synthesized pieces with identified and predictable functionality across different platforms. Exemplifying this strategy, the Registry of Standard Biological Parts, or "BioBricks,"TM physically houses an open catalog of standardized DNA parts that encode basic biological functions and can be easily combined and exchanged among different devices and laboratories.[21] These standardized parts are made available to the public free of charge to further research in this field, and they are central to the annual International Genetically Engineered Machine (iGEM) student competition.

Defining Synthetic Biology

Despite these historical antecedents and complementary methodologies, providing a single definition for synthetic biology is a challenge even to those active in the field. Synthetic biology has attracted interest and investment from a range of different specialties. Biologists, chemists, engineers, and others bring their collective knowledge and expertise to this inherently interdisciplinary science. For this reason, synthetic biology may be viewed from various perspectives, which together help to explain its utility and versatility. A common thread is that synthetic biology is a scientific discipline that relies on chemically synthesized DNA, along with standardized and automatable processes, to address human needs by the creation of organisms with novel or enhanced characteristics or traits. To a biologist, synthetic biology is a window through which to understand how living things operate. It provides a direct and compelling means to test, through sequencing, modeling, and reproduction, our current understanding of the life sciences. The ability to model and manipulate living systems using synthetic biology is yielding new knowledge that will better define the functions of genes and physiological systems. In addition to advancing basic science, synthetic biology has important potential applications for medicine, including the design of safe and effective vaccines and targeted approaches to detect and cure diseases like cancer.

iGEM

The iGEM competition resembles a giant science fair for budding synthetic biologists. iGEM is a global synthetic biology competition involving mostly undergraduate students, although non-synthetic biology faculty, and high school students also participate. At the heart of the competition is BioBricks, a repository of standard DNA parts. Several months before the actual competition, competing teams receive a kit of DNA parts. Working at their own schools over a summer, teams design and build synthetic systems that operate in living cells. Examples of recent projects include an arsenic biosensor, wintergreen-scented bacteria, and color-coded microbes. Teams earn medals in a range of categories. Among the more popular of these is "human factors." Here, competitors win points for innovations that directly affect how people work together. Beyond building biological systems, the broader goals of iGEM include growing and supporting a community of science guided by social norms.

From the perspective of a chemist, synthetic biology is a tool for manufacturing novel molecules and molecular systems for various uses. Scientists have used synthetic biology to directly manipulate chemical reactions in living systems, for example, in hopes of making medicines quickly and inexpensively.[22]

They have also produced, on a small scale, novel biofuels that can harness energy from plants and the sun.[23] Collectively, these methods could reduce the use and deleterious effects of hazardous chemicals and petroleum-based products.

Synthetic biology viewed through an engineering lens is an opportunity to apply the techniques and tools of engineering to complex living organisms. Many aspects of engineering are based on the principle of standardization, which enables the reliable production of useful commodities. Engineers working in the field of synthetic biology hope to bring a similar level of standardization, predictability, and reproducibility to biology.

Examples of engineered biological systems currently under study include synthetic systems that perform sophisticated medical functions—measuring components in body fluids and adjusting them through targeted administration of therapies—as well as biologically engineered "microcleansers" that can clean up oil spills or other forms of industrial waste.[24]

Is Synthetic Biology New?

The answer to this question is complex. Some scientists see synthetic biology as a revolutionary and qualitatively new field of science.[25] Others see current developments in the field as incremental advances in the decades-long growth of molecular biology, genetic engineering, and microbiology.[26] The term synthetic biology itself was first used as early as 1974 by Waclaw Szybalski who saw molecular biology's promise evolving from description to manipulation of genetic systems, heralding a new era of synthetic biology.[27]

One characteristic that distinguishes the synthetic biology of today from the molecular biology of years past is the significant role played by standardized parts, computers, and automation, accelerating a trend prevalent throughout biotechnology.

Companion fields like nanotechnology and biomedical imaging share a reliance on automation and reusable, standardized parts.

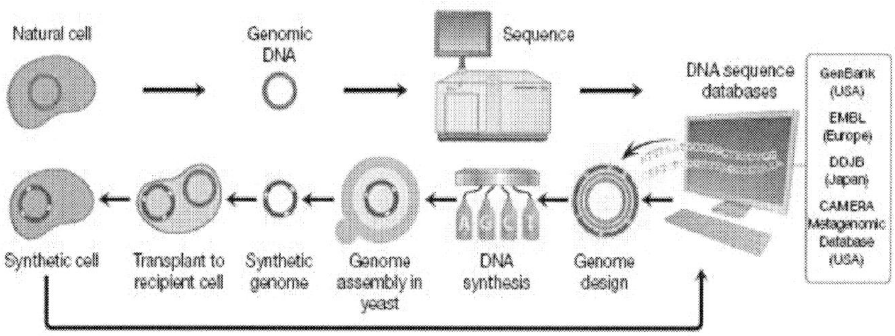

Figure 6. Overview of one process using synthetic biology techniques to produce synthetic cells. (Courtesy of J. Craig Venter Institute).

Recent technological advances and economic efficiencies in DNA synthesis and sequencing permit synthetic biologists to make, move, and manipulate DNA on a much larger scale than was possible only a few years ago.

In contrast to conventional research in biology, the quest for predictable functions and standardization lies at the heart of synthetic biology. In this way, the field reflects the influence of engineering on its development.

The Future of Synthetic Biology

Synthetic biology holds great promise as a route to develop novel applications for medicine, agriculture, energy, and other industries. For example, the future may hold microorganisms that are "tailor-made for production of a specific chemical from a specific starting material . . . ."[28] Few of these potential products are anticipated immediately, however, and considerable technical and intellectual challenges remain.

Building a single cell from parts in the laboratory is a vastly different challenge than building an organism that interacts effectively and predictably in nature.[29] The design of synthetic or artificial organisms that can survive in natural environments is likely to be more challenging and unpredictable than doing so in a controlled setting.[30] It is extremely difficult to anticipate with confidence how a synthetic organism will react to and interact with a novel natural environment, adding to concerns about the risks of some applications of this field (see Chapter 3 for further discussion of applications).

Complexity and variation are linked. They both reflect the fact that DNA alone is not sufficient to create the biological functions necessary for the creation of biofuels, vaccines, soil sensors, or any desired product of synthetic biology. DNA can only function if it exists within an environment that provides the cellular components such as ribosomes, proteins, and other structures necessary to read, translate, and implement its genetic code. How any specific DNA sequence functions in a cell is also dependent on secondary modifications in its structure (though methylation) or folding pattern (through changes in histone proteins) that can promote or inhibit the transcription of genes, an area known as *epigenetics*. Much is still unknown regarding the interactions between and within cells, actual or "artificial," as well as between cells and their environments.

Currently, the behavior of synthetic biological systems remains unpredictable.[31] Function cannot typically be accurately predicted based on DNA sequence alone or by the shape and other characteristics of the proteins and the biological systems for which it codes.[32] Also unknown is how synthetic biological systems will evolve. In most cases, biological systems that have been engineered by scientists quickly revert to "wild type" (i.e., evolve to lose their engineered function rather than gain a new one).[33] Although this notion may be reassuring, it does not rule out the possibility that systems might evolve in unpredictable and harmful ways, particularly if released outside the laboratory.

The potential promise of synthetic biology is immense. Research in synthetic biology has led to the development of genetic circuits and modules with predictable behavior, creation of novel combinations of cells in the laboratory that behave synergistically, and ever-expanding DNA construction capabilities.[34] The field, however, is young. Our understanding of complexity and variation in natural and synthetic parts and systems is far from complete, and the technical tools and skills required for large-scale synthesis and production continue to be refined. If carefully nurtured and guided, however, synthetic biology may provide an opportunity to integrate engineering and the biological sciences into the living world, with potential benefits to national and international security, food and energy supply, public health, and economic well-being.

APPLICATIONS, BENEFITS, AND RISKS

Synthetic biology offers opportunities to apply biological and engineering principles to benefit humankind in unprecedented ways. Clean energy sources, targeted medicines and more efficient vaccine production, new chemicals, environmental cleansers, and hardy crops are some of the potential applications of this burgeoning field of science. While most of the fruits of synthetic biology remain in early stages of development, some applications are expected to come to market within a few years.[1] Success in these research efforts will yield new jobs as novel products and product streams develop. The pace of acceleration of synthetic biology is likely to increase dramatically in the years ahead.

Despite its promise, synthetic biology raises concerns about risks to human health, the environment, and biosecurity. Some of these potential harms include unanticipated adverse human health effects, negative environmental effects (anticipated or unanticipated) from field release and dual-use concerns when research undertaken for "legitimate scientific purpose...may be misused to pose a biologic threat to public health and/or national security."[2]

This chapter provides an overview of the potential applications, benefits, and risks of synthetic biology. Because renewable energy is expected to yield the first large-scale commercial products of synthetic biology, the Commission discusses this area first. Next, the Commission reviews potential health applications and benefits. Many products remain in research and development, but a few are nearing commercialization. Finally, the Commission provides a summary of potential agricultural, environmental, and biosecurity applications of synthetic biology, all of which are in more preliminary stages of development. Within these discussions the potential health, security, and other risks are examined, as well as anticipated technical challenges.

Renewable Energy Applications of Synthetic Biology

In general, biofuels are renewable energy sources derived from biomass, which includes material derived from plants, animals, and organic waste. Several methods can be used to harvest energy from biomass, including burning, chemical treatment, or biodegradation using the metabolic power of microorganisms. Processing biomass into biofuels or electricity through more complex chemical and biochemical reactions, as opposed to simple combustion, limits environmental impact by minimizing the production of waste and decreasing net greenhouse emissions. Current practices for farming biomass for energy use employ a range of biological sources including grains, grasses, oil seed crops, trees, sugar, and corn.

Ethanol is the most common biofuel worldwide. It is produced mainly from corn or sugar cane. Biodiesel, another currently used biofuel, is made from vegetable oils, animal fats, or recycled restaurant grease. There are challenges to widespread commercial development of either of these fuels. For ethanol production, challenges include inefficiencies and energy costs for production, as well as concerns about the volume of plant sources needed and possible collateral impact on food prices. Biodiesel also involves significant energy costs for production.

Promise and Potential Benefits

Biofuels and related products produced through synthetic biology offer the potential to reduce global dependence on fossil fuel, cut harmful emissions, and minimize economic and political volatility surrounding fossil fuel reserves. Some biofuels produced with synthetic biology processes are expected to be available commercially within the next few years. Other research may not yield commercial products for a decade or more.

The various synthetic biology alternatives to current biofuel production methods include producing cellulosic ethanol (derived from cell walls rather than corn) and manufacturing other bioalcohols with synthetically manipulated biomass. Biofuel can also be produced from modified algae that use the natural process of photosynthesis to manufacture bio-oils, such as biodiesel, more easily than current chemical processes.[3]

The biochemical conversion of biomass into energy involves chemical reactions performed by biological systems. Enzymes in microorganisms such as bacteria break down biological materials into their component parts, from which energy can be extracted more easily. Perhaps the simplest example of biochemical conversion is a backyard composting bin, in which microorganisms gradually degrade vegetation in the presence of oxygen. As is apparent from the surge of warm air that emerges upon opening the lid of the bin, this form of bioconversion is an energy-yielding process.

Synthetic biologists aim to improve the speed and efficiency of converting biomass into advanced, second- or third-generation biofuels with cleaner and more favorable energy-usage profiles.[4] This challenge may be met by creating "super-fermenting" yeast and bacteria through synthetic biology. These organisms have the potential to boost the power and potential of current industrially used microorganisms by means of new or altered genes. Synthetic biology also offers new biomass sources, or feedstocks, that are more efficient, reliable, low-cost, and scalable than current sources. These include forest and agriculture residues, some grasses, algae, oilseeds, and potentially sewage.[5]

Aside from biofuels, synthetic biology may also play an important environmental role by harnessing energy in novel, cleaner ways than traditional non-renewable energy production processes. Large global reserves of hydrocarbons, such as oil, gas, shale, and oil sands, might be leveraged with synthetic biology tools. Coal bed methane, for example, is a globally available source of natural gas. Its reserves are vast and largely untapped. Synthetic biology research is underway to harvest this methane through microbial digestion and other processes.[6]

Bioalcohols

Unlike ethanol derived from corn or sugar cane, cellulosic ethanol is made from cellulose fibers, a major component in the cell walls of all plants. Processing plant biomass not used for food, for example, waste corn stalks, straws, grass clippings, prairie grasses, and wood chips, could reduce economic and other pressures imposed by relying on corn for ethanol. However, cellulosic ethanol is a relatively low-yield bioalcohol and, like ethanol fuel derived from more conventional chemistries, still tends to corrode storage and transport equipment.

New Product Pipeline: Bioalcohols

Amyris (Emeryville, California) is using a synthetic biology platform to convert sugar into a range of products, including yeast-derived cellulosic alcohol fuel. The oil-based fuel is harvested in a similar fashion to the technique used by the Joint Bioenergy Institute (akin to separating cream from milk).[7]

British Petroleum and **DuPont** created a partnership to develop, produce, and market biobutanol.[8]

Gevo (Englewood, Colorado) genetically engineered bacteria to make biobutanol, a promising new biofuel. It also successfully converted cellulosic biomass into iso-biobutanol and converted the fuel into jet fuel.[9]

Global Bioenergies (Evry, France) created yeast and bacteria with the capacity to transform sugar into hydrocarbons chemically identical to those distilled from oil. Bio-isobutane is the targeted end product; this hydrocarbon gas can be converted into high-octane gasoline.[10]

The U.S. Department of Energy's **Joint Bioenergy Institute** (Emeryville, California) is using synthetic biology to biodegrade plant biomass into biodiesel, which is skimmed off the top of a fermentation broth.[11]

LS9, Inc. (South San Francisco, California) developed the UltraCleanTM product line that employs synthetic biology to produce its DesignerMicrobesTM. These microorganisms use sugar cane or cellulosic biomass to create high-energy transportation fuels.[12]

The various commercial products and products presented and described in this report are intended to provide examples of current projects, not to endorse any particular entities.

A potentially more promising bioalcohol made by synthetic biology and used for energy production is butanol. Like ethanol, butanol is produced by the fermentation of sugars and starches or through the breakdown of cellulose. The crude product is then refined to make usable fuel. A particular advantage of butanol (and a similar biofuel called isobutanol) is that it can be used directly in a traditional gasoline-powered engine. It also has a relatively high-energy density, resulting in better gas mileage than ethanol.[13] Some bacteria have the built-in enzymes to manufacture butanol, but the natural process is not very fast or high-yield. Synthetic biologists have engineered the easy-tomanipulate bacterium *E. coli* to improve this bacterial biochemical reaction to make butanol more industrially useful.[14]

Photosynthetic Algae

Another tool for creating biofuels via synthetic biology is through the use of photosynthetic algae. Algae are low-input, high-yield feedstocks that, under experimental conditions, produce substantially more energy per acre than land crops such as corn or soybeans.[15]

New Product Pipeline: Photosynthetic Algae

Aurora Algae (Alameda, California and Florida) is growing algae in open-pond systems consisting of readily available seawater. The pilot facility in Florida produces approximately three tons of algal biomass per year, with the ultimate goal of producing 40,000 tons of algal biomass per year.[18]

Joule (Cambridge, Massachusetts) engineers algae to make and secrete liquid hydrocarbons, bioethanol, and other fuel materials from sunlight and waste carbon dioxide (the sole feedstock) in a single-step, continuous process. Pilot operations are currently underway, with commercial development slated for 2012.[19]

Solazyme (South San Francisco, California) uses photosynthetic algae to produce an oil-based fuel, Soladiesel®, at industrial manufacturing scale with production capabilities currently in the tens of thousands of gallons. In July 2010, Solazyme delivered 1,500 gallons of algal-derived jet fuel to the Navy.[20]

Synthetic Genomics Inc. (La Jolla, California) engineered algal strains to create a biocrude oil that can be used as a feedstock in refineries, using a continuous biomanufacturing process that sidesteps the intermittent cycle of growing and harvesting. In July 2009, Synthetic Genomics entered into a $600 million multi-year agreement with ExxonMobil.[21]

The various commercial products and products presented and described in this report are intended to provide examples of current projects, not to endorse any particular entities.

To create biofuel from algae, the cells are grown, harvested, and treated chemically or thermally to recover the oil content inside algal cells, the so-called "bio-oil." While experimental yields have not yet been duplicated on a commercial scale, an alternative strategy currently under development with synthetic biology is engineering algal cells to secrete oil continuously through their cell walls and thereby increase yield. This time-saving step may support large-scale industrial operations in the near future.[16]

Proponents of farming algae note that it is biodegradable and therefore relatively harmless to the environment if spilled. Algae can also be grown on land and in water that is otherwise unsuitable for crops and food production. Making bio-oils using algae is expected to be less polluting and more efficient than converting vegetable oils or animal fats into biofuel.[17]

Through its capacity to consume carbon dioxide, algae offer the added benefit of mitigating greenhouse gas emissions. unlike ethanol, algae-derived biooils, such as gasoline, diesel, and jet fuels, have been found to have very similar physical and chemical properties in comparison to currently used petroleum-based products, suggesting that these fuels are likely to be compatible with current transportation technologies and infrastructure.

Hydrogen Fuel

Hydrogen fuel is an additional area of focus for commercial applications of synthetic biology. Hydrogen is a highly desirable fuel source because it is clean-burning, producing water as a by-product. Hydrogen also has the second highest energy density per unit of weight of any known fuel.[22]

Several possible routes to generate biohydrogen are under investigation. One method uses engineered *E. coli* as a host organism to produce hydrogen in addition to other biofuels.[23] Engineered algae are also being examined as sources of biohydrogen.[24] Finally, and perhaps most promisingly, researchers are investigating ways to produce high yields of hydrogen using starch and water via a synthetic enzymatic pathway.[25] The latter system is particularly attractive, as it may enable sugar to be converted into hydrogen fuel inside a vehicle itself. This would mitigate the problem of storage that exists today, as hydrogen takes up inordinate amounts of space at regular atmospheric pressure and compression of the gas requires energy and makes storage both difficult and dangerous.[26]

The synthetic processes being explored, if successful, will differ markedly from the current method of producing hydrogen fuel, which involves converting natural gas using steam. Natural gas techniques are costly, inefficient, and heavily reliant on fossil fuels. The synthetic biology-driven process is expected to cost significantly less while providing substantially higher yields, though research remains early in the developmental pipeline.

Risks and Potential Harms

Synthetic biology offers many potential methods to improve energy production and reduce costs, which deservedly generate attention and enthusiasm. A full assessment of these promising activities requires comparable attention to the current limitations, challenges, and anticipated risks or harms. This assessment is particularly important at this time because renewable energy applications may be the first synthetic biology products to come to market.

Contamination by accidental or intentional release of organisms developed with synthetic biology is among the principal anticipated risks. unlike synthetically produced chemicals, which generally have well-defined and predictable qualities, biological organisms may be more difficult to control. unmanaged release could, in theory, lead to undesired cross-breeding with other organisms, uncontrolled proliferation, crowding out of existing species, and threats to biodiversity.[27]

Consider biofuel production systems that employ synthetic biology and pond-grown algae. One hypothetical, worst-case scenario is a newly engineered type of high-yielding blue-green algae cultivated for biofuel production unintentionally leaking from outdoor ponds and out-competing native algal growth.[28] A durable synthetic biology-derived organism might then spread to natural waterways, where it may thrive, displace other species, and rob the ecosystem of vital nutrients, with negative consequences for the environment.

This scenario is theoretical. Considering it and developing appropriate precautions is nevertheless appropriate because of the rapid development of synthetic biology-generated photosynthetic algae for fuel production and the uncertain nature of the harm that may arise from accidental release. One of the advantages of synthetic biology is that many of the tools being developed include strategies to remediate such risks. Some of the approaches proposed include the engineering of so-called “terminator” genes or “suicide” switches that can be inserted into organisms, precluding them from reproducing or surviving outside of a laboratory or other controlled setting in the absence of unique chemical conditions.[29] Some are clearly sufficient to neutralize the risk of release, and others require further study as synthetic biology progresses.

Another risk in the energy sector is harm to ecosystems from the required dedication of land and other natural resources to production of biomass as feedstock for biofuels. If large areas of land were to be dedicated to biofuel development, this could put new and intense pressures on land, potentially affecting food production, communities, and current ecosystems. Because these applications of synthetic biology are still young, the impact of biofuel production on land use remains unknown. Some argue that efforts to develop and grow additional cellulosic biofuel will dramatically change and adversely impact the way land is used in the United States and abroad.[30] Others suggest that biofuel production can proceed safely with only minor adjustments in current land use practices.[31] Existing biodiverse prairie and meadow grasses may actually enhance the growth of feedstock for second-generation biofuels.[32] On balance, many anticipate the potential efficiencies and attendant reduction in reliance on fossil fuels offered by energy production using synthetic biology would offset anticipated risks to the environmental ecosystem as it exists today. But considerable uncertainty remains.

Health Applications of Synthetic Biology

Synthetic biology has the opportunity to advance human health in a variety of ways. Improved production of drugs and vaccines, advanced mechanisms for personalized medicine, and novel, programmable drugs and devices for prevention and healing are among a few of the expected achievements.

Promise and Potential Benefits

There is a long tradition of employing plants and other biological organisms to detect and cure human disease. Genetic engineering technology has been used for more than three decades in medicine to engineer bacteria with the ability to produce commercially relevant molecules like insulin and vaccines for hepatitis B virus and human papillomavirus.[33] Synthetic biology applications related to health build on this history, but most remain early in the research and development pipeline. The quick pace of biomedical research in general, and synthetic biology research in particular, suggests that this could change soon. This research is being conducted at universities and biotechnology or synthetic biology companies in the united States and overseas.[34]

Medicines

Synthetic biologists have refined a chemical technique called metabolic engineering to enhance the production of medicines. Through this process, scientists alter an organism's metabolic pathways—the series of chemical reactions that enable the organism to function at the cellular or organism level—in order to better understand and manage how those pathways work. They can redesign these pathways to produce novel products or augment the production of current products, like drugs.

Synthetic biology can also be used to engineer molecules and cells that express proteins or pathways responsible for human disease. At some point these products may be used in efficient, large-scale screening methods to identify novel drugs for disease treatment or prevention. One well-known example of synthetic biology in medicine is the re-engineering of a microorganism to make the antimalarial drug artemisinin more cheaply and efficiently. Malaria affects approximately two to three hundred million people each year and results in between 700,000-1,000,000 deaths, largely among young children in sub-Saharan Africa.[35] Artemisinin is a naturally occurring chemical derived from the plant artemesia, or sweet wormwood. It is an effective malaria treatment, but is difficult to obtain due to limitations on plant yield and high production costs. To address this problem, synthetic biologists at the University of California genetically engineered *E. coli* bacteria to produce a high volume precursor that can be chemically converted to artemisinin.[36] This semi-synthetic artemisinin is being developed today by the pharmaceutical company Sanofi-Aventis in collaboration with the California researchers and the Institute for OneWorld Health. If successful, these efforts should substantially reduce the drug's production cost and increase and stabilize world supply. Full-scale production is expected to begin shortly, with marketing expected in 2012.[37]

> "Making a few micrograms of artemisinin would have been a neat scientific trick," said Dr. Jay Keasling, whose laboratory originally developed the synthetic biological concept for making artemisinin. "But it doesn't do anybody in Africa any good if all we can do is a cool experiment in a Berkeley lab.
> We needed to make it on an industrial scale."[38]

Vaccines

Synthetic biology techniques are also being studied and used to accelerate the development of vaccines. Influenza vaccine production is among the key areas of focus. To develop a vaccine, one first needs to identify the virus strain, with its unique genetic code, against which the vaccine will be used. Synthetic biology tools, including rapid, inexpensive DNA sequencing combined with computer modeling, may streamline production time by accelerating this first step. One industry group is developing a "bank" of synthetically created seed viruses for influenza vaccines that it hopes will enable more rapid vaccine production by reducing virus identification time.[39] DNA-based vaccines created "on-the-spot" to match actual, circulating viral genetic material may be a more efficient process for producing vaccine seed stock in the future.[40] However, these strategies are preliminary and may prove no more efficient or effective than conventional reverse engineering techniques. More research and experience is needed.

Advancing Basic Biology and Personalized Medicine

Twenty years ago, cloning, or replicating, a single gene was enormously time consuming. Today, such a task can be done in minutes by a machine, a development that has fueled rapid advances in synthetic biology. The ability to easily manufacture and manipulate DNA in the laboratory has enhanced scientists' productivity and opened new directions for scientific exploration. Researchers see great potential for synthetic biology to advance knowledge of fundamental biological principles. Expanding the DNA "alphabet" beyond its traditional four nucleotides—A, C, G, and T—to include non-naturally occurring nucleotides also gives synthetic biologists more flexibility in studying, detecting, and treating disease. For example, scientists recently used polymerase chain reaction (PCR) with novel nucleotides, a process that increases DNA's information potential and thus enables the manufacture of proteins with new properties.[41] To this end, researchers have already developed diagnostic tests using these DNA nucleotides to screen for human immunodeficiency virus, cystic fibrosis, and other diseases.[42]

In general, personalized medicine aims to apply the science of genomics to develop individually tailored, and thereby more effective, approaches to disease prevention and health care.[43] Synthetic biology offers useful strategies for advancing this goal. Many current cancer treatments focus on non-selective cell killing or on delivery to specific tissues. A growing body of knowledge supporting a molecular classification of tumors may facilitate the development of specifically designed detection devices matched to individual tumors. A synthetic biology approach currently under study is a cancer treatment that focuses on up to six cellular identifiers rather than one, effectively enabling the treatment to be targeted more carefully and precisely toward the cells intended to be killed, while sparing healthy ones.[44]

Custom protein and biological circuit design may eventually enable the delivery of "smart proteins" or programmed cells that self-assemble at disease sites. Similarly, synthetic organisms could be developed to create a trigger to deliver or withhold treatment depending upon a local disease environment (such as low levels of oxygen) and provide targeted killing of cancer cells.[45] These and other novel approaches to tailored disease treatment may substantially improve outcomes and reduce the costs and burden of disease across the population.

While the benefits of synthetic biology to health care may prove monumental, significant hurdles remain. With the exception of semi-synthetic artemisinin and potential, near-term improvements in vaccine design, most of the anticipated health benefits of synthetic biology remain in the preliminary research stage. We are unlikely to see commercial applications from much of the biomedically oriented synthetic biology research for many years, although the pace of discovery is unpredictable.

Risks and Potential Harms

In addition to practical challenges, biomedical applications of synthetic biology raise potential risks for humans and the environment that are, in part, similar to those identified in the biofuels discussion and those commonly understood within the biomedical or greater engineering research communities today. Human health risks may arise from adverse effects of intentional or inadvertent release of the organisms engineered using synthetic biology. Infectious diseases may be transmitted to laboratory workers after needle sticks or to family members following airborne transmission of disease agents manipulated using synthetic biology techniques. Risks may also accrue to the wider human community or the environment if organisms proliferate without adequate means to limit reproduction.

Similarly, novel organisms developed with synthetic biology to treat illness may trigger unanticipated adverse effects in patients. The use of cell therapies of bacterial, or potentially, mixed microbial origin may cause infections or unexpected immune responses. New organisms developed with the emerging technology of synthetic biology may pose unusual, if not unprecedented, risks resulting from their potential as biological organisms to reproduce or evolve.

Many of these risks are qualitatively similar to the risks that arise in horticultural biomedical and biotechnology research. There are well-established mechanisms in place to identify and manage future risks (see Chapter 4). Additionally, as with energy applications, internal mechanisms to reliably contain function and reduce or eliminate these risks are being developed. "Biological isolation," which is also termed "biosafety engineering," aims to build in molecular "brakes" or "seatbelts" that restrain growth or replication of partially or fully synthetic organisms.[46] Synthetic organisms can be engineered to be contained physically or temporally. Additional data are needed to assess how well biologically engineered safeguards, such as "kill switches" that activate after a defined number of generations, will work.

Agricultural, Food, and Environmental Applications of Synthetic Biology

Synthetic biology may also help to shift, if not substantially mitigate, some of the existing threats to our global food supply and environmental health. These potential benefits are in some ways more preliminary than the expectations for energy and health, but research and development in these fields are well underway.

Promise and Potential Benefits

In agriculture, efforts to manipulate crops and breed animals for specific purposes are not new. Many traditional farming practices, from plant breeding to animal husbandry, aim to direct evolution to achieve desired outcomes. use of recombinant DNA technology, cloning, and other biotechnology tools have enhanced these practices. Taking these activities one step further, synthetic biologists are experimenting with high-yield and disease-resistant plant feedstocks that can be supplemented with efficient and environmentally friendly microorganisms to minimize water use and replace chemical fertilizers.[47] Researchers are altering the properties of plants through methods that combine metabolic components from various organisms in order to gain nutritional benefits, such as higher levels of food-grade protein.[48]

New Product Pipeline: Crop Enhancement And Pollution Control

A synthetic biology-produced Pyrethium-grown compound may find use as natural insecticide.[51]

Synthetic biology-produced DNA sensors may be able to perform a range of roles, including detecting food spoilage and monitoring soil nutrition.[52]

Synthetic biology technology has been proposed to control biodegradation of a range of sources including toxic chemical pollutants such as industrial coolants, solvents, explosives, and residues from burning oil, coal, and tar.[53]

Efforts to remove waste using biological means date to at least 1972, when a researcher at General Electric applied for a patent on a form of *Pseudomonas* bacteria genetically engineered to digest oil slicks.[49] Environmental applications of synthetic biology are generally targeted to pollution control and ecological protection. The impact of naturally occurring oil-devouring microorganisms at the site of the 2010 oil spill off the U.S. Gulf Coast, for example, demonstrated how these organisms could reduce some types of pollution.[50] Synthetic biologists are eager to understand and direct these biological capabilities, or even enhance them, to respond to existing and future waste generated by human activities.

Other environmentally relevant examples of synthetic biology applications include laboratory-constructed microbial consortia, known as synthetic biofilms, which are being developed for use as environmental biosensors. These sensors could be used, for example, to monitor soil for nutrient quality or signs of environmental degradation. The design of biological "wetting agents," or biosurfactants, could increase the efficiency of bioremediation efforts and minimize the extent of damage from pollutants.[54] Biosurfactants are naturally produced by bacteria, yeasts, or fungi and are environmentally friendly in freshwater, marine, and terrestrial ecosystems. Synthetic biology may offer the ability to enhance the features of microbially produced biosurfactants to tailor them to specific spills or otherwise polluted areas.

Risks and Potential Harms

Synthetic biology applications in the context of agriculture, food, and the environment raise concerns broadly similar to those raised about genetic engineering in the past and those discussed above with respect to safety, resource management, and biodiversity. In brief, these risks include harms to humans, plants, or animals from, for example:[55]

- uncontrolled environmental escape or release and attendant disruption to ecosystems,
- new or sturdier pests—animal or plant—that may be difficult to control, and
- increased pesticide resistance and growth of invasive species.

As in the discussion of energy and health applications, the risks may be assessed and managed through existing protections long in use for biomedical and greater engineering research. Synthetic biology applications in the context of agriculture, food, and the environment may require more targeted efforts, however, including use of inbred checks, such as "suicide genes" or "kill switches" to ensure that they cannot propagate unintentionally.

Many potential applications of synthetic biology go well beyond the genetic engineering practiced throughout the biotechnology industry today. In the future, the field may be capable of creating entirely new organisms and systems previously unseen in the world today. Synthetic biology's critics and proponents alike worry that creating new organisms that have uncertain or unpredictable functions, interactions, and properties could affect ecosystems and other species in unknown and adverse ways. The associated risks of escape and contamination may be extremely difficult to assess in advance, as such novel entities may have neither an evolutionary nor an ecological history.[56]

Countering these concerns, at least somewhat, is experience showing that synthetic cells and systems in research settings have tended to be short-lived by comparison to those that have evolved in nature. Scientists have observed that synthetic organisms allowed to develop in the laboratory have consistently evolved toward nonfunctionality.[57] These are encouraging preliminary findings, but they do not eliminate the need for precautions in the event that a future synthetic organism behaves differently than expected outside of the contained laboratory setting.

Another concern related to synthetic biology's impact on natural systems— crops grown for either biofuel or food consumption—is the broader effect on how society views and protects biodiversity. Does a chemically synthesized organism increase or decrease biodiversity, as measured by traditional taxonomy-based classification schemes? This concept becomes important in policy discussions pertaining to the use and potential abuse of land and other natural resources.

Biosecurity

Generally, the term "biosecurity" refers to the efforts needed to prevent misuse or mishandling of biological agents and organisms with the intent to do harm. The National Science Advisory Board for Biosecurity (NSABB), an independent federal advisory committee charged with advising the U.S. government on biosecurity issues and "dual use" research—that which may be used for either good or ill—defines the term as follows: "[b]iosecurity refers to the protection, control of, and accountability for high-consequence biological agents and toxins, and critical relevant biological materials and information, to prevent unauthorized possession, loss, theft, misuse, diversion, or intentional release."[58]

Unlike applications and potential applications of synthetic biology in the energy, health, agricultural, and environmental sectors, possible benefits in the biosecurity arena have not garnered significant public attention. Nor have they received comparable investment from academia, industry, or the government. It is nonetheless easy to anticipate some potential benefits.

Synthetic biology may enhance biosecurity by enabling researchers to identify biological agents of concern that may be developed synthetically or semi-synthetically. In the same way that the J. Craig Venter Institute "branded" the bacterium it synthesized this year with traceable information in the organism's genetic code, researchers may uniquely tag the genetic code of new organisms that they develop. When combined with other measures to ensure biosecurity, this tagging process may provide an additional and effective deterrent to malicious use.

Similarly, biosecurity may be improved using the techniques discussed above for applications in energy, human health, agriculture, and the environment. As noted, "suicide" genes or terminator technologies built into the genome of a new organism to inhibit growth or survival outside of a contained environment may offer particularly effective means to counter biosecurity threats. Related tools could be crafted to ensure organism death in the face of particular chemicals or contexts. uncertainties remain, however, with regard to the effectiveness of such strategies.

Concerns about dual use or intentional misuse of synthetic biology to do harm are among the most prominent critiques of this emerging technology. One of the most widely voiced risks attributed to synthetic biology is that it may be used, in the wrong hands, to intentionally create harmful organisms for bioterrorism. Recent examples of virus reconstruction using traditional recombinant DNA techniques fuel these concerns. These examples include the laboratory creation of infectious polio virus, the mycoplasma genome, and the 1918 strain of influenza virus.[59]

Frequently lost in these discussions about synthetic biology risks is recognition that DNA alone is not sufficient to create an independently functioning biological entity, such as a disease-causing virus that could spread. Despite the relative ease of access to known DNA sequences through public databases like GenBank[60] (an annotated collection of all publicly available genetic sequences), and equivalent databases across the globe, most experts in the scientific community agree that mere knowledge of a viral genome is far from sufficient to be able to re-constitute it or create a disease-forming pathogen. Rather, one must have an appropriate host and conditions for a virus to grow. Few individuals or groups today have the financial means or the technical skills to accomplish such ends, even when scientifically feasible. As the many technical challenges in synthetic biology affirm, it is not yet possible to craft functioning biological organisms from synthesized genomic material alone.

Risks and Potential Harms

With regard to biosecurity risks arising from synthetic biology, NSABB has twice issued reports and made recommendations to the federal government— first in 2006 and again in 2010.[61] In 2006, the group focused on synthesis of select agents and toxins, which are defined in law as certain infectious components of identified "select agent viruses," meaning those that the U.S. government has found to pose a severe threat to human health.[62] Following a review of the science at that time, the group made specific recommendations to reduce biosecurity risks, many of which the United States has since implemented, such as the establishment of a screening infrastructure for genetic sequence providers and others.[63]

NSABB's report "Addressing Biosafety Concerns Related to Synthetic Biology," issued in April 2010, offered four specific recommendations to ensure biosecurity in the current field of synthetic biology:

- Synthetic biology should be subject to institutional review and oversight since some aspects of this field pose biosecurity risks.
- Oversight of dual use research should extend beyond the boundaries of life sciences and academia.
- Outreach and education strategies should be developed that address dual use research issues and engage the research communities that are most likely to undertake work under the umbrella of synthetic biology.
- The U.S. government should include advances in synthetic biology and understanding of virulence/pathogenicity in efforts to monitor new scientific fi ndings and technologies.

These recommendations reflect an attempt to balance the considerable potential benefits of synthetic biology with the risks resulting from intentional or unintentional misuse of this technology and its products. Noticeably absent were recommendations to restrict access to genetic sequences separate from those components of Select Agents and toxins already limited by the U.S. Select Agent regulations (see Chapter 4). In large part, this determination appears to reflect the fact, as noted, that sequences alone will not yield, nor often be sufficient to predict, functions.

NSABB's work is not unique. Many experts and interested groups in the united States and abroad have recently devoted considerable time and energy to evaluating the biosecurity risks of advancing synthetic biology practices.[64] This still-young field benefits from a clear consensus among scientists and policymakers that biosecurity risks, while perhaps overstated by some, nevertheless are serious and warrant ongoing and proactive re-examination as technical capacity evolves. The tools used to mitigate these risks may also be the tools to mitigate environmental, health, and other potential risks. The tools to address risk depend on an expanding scientific knowledge base as much as potential benefits do.

OVERSIGHT

A wide array of existing federal laws and regulations apply to the emerging field of synthetic biology. The scope of federal authority depends on whether the activity involves research or production; whether federal funds are involved; the nature of the application (e.g., to generate drugs, food, cosmetics, or fuels); and whether the product is subject to national security or export controls. Applicable also are local institutional, municipal, and state requirements, many of which focus on safety and security.

This chapter presents a brief overview of the components of the u.S. oversight system as it relates to synthetic biology. It is intended to be a descriptive summary of the major regulatory laws and agencies without Commission recommendations or opinion (presented in Chapter 5). It focuses on the exclusive, as well as shared and overlapping, federal authorities governing research, development, and commercialization. Generally, synthetic biology is treated like other comparable areas of science and technology, and the federal government relies, in part, on local institutional-level oversight to identify and reduce risks.

The government's initial efforts at oversight of genetic engineering activities arose in the mid-1970s and focused, consistent with the state of the science at the time, on laboratory-contained research.[1] When the first genetically engineered organisms were being considered for field testing in the mid-1980s, the u.S. government issued a trans-agency guidance document, called "The Coordinated Framework," for regulating the research and development of biotechnology products. Fundamentally, the policy calls for the government to regulate genetically engineered products through existing legal frameworks established for products developed without genetic engineering. For example, drugs developed by means of genetic engineering are regulated under the pre-market review and approval standards of the Food and Drug Administration (FDA) for new drugs.[2] The key to this policy, reflected in regulations across the government, is its focus on risk rather than methodology. Regulation is predicated on a risk-benefit assessment of the characteristics of the final product (i.e., its intrinsic characteristics and features), not the method by which it is made.[3] Products presenting higher risks or greater uncertainty are subject to higher degrees of oversight. This approach enables existing agencies and regulations to serve, with revisions in current rules as technology evolves, as the oversight framework for emerging biotechnology. Periodic reassessment, ideally through an ongoing process of open public dialogue, is required as new knowledge and new understanding of risks emerge. The Coordinated Framework's standards continue to drive the federal government's approach to oversight of biotechnology, including synthetic biology.

Through this system, some oversight protections apply broadly to anyone working with specific organisms or creating certain environmental effects. Other oversight is more narrowly focused, applying exclusively, for example, to researchers or the research setting. Regulatory programs of the U.S. Department of Agriculture (USDA) or FDA apply case-by-case to particular goods like food or drugs. USDA regulations govern also the interstate movement of certain infectious agents, agricultural pathogens, and pests. The Environmental

Protection Agency (EPA) regulates the safety of new chemicals not addressed by other statutes, including industrial chemicals and pesticides, and oversees emergency management programs for the clean up of environmental hazards. The Occupational Safety and Health Administration (OSHA), Department of Transportation (DOT), and Department of Commerce (DOC) play roles as well, setting safety standards respectively for the workplace, interstate transfer of infectious agents, and export of disease-causing organisms or knowledge and technologies that may pose security risks.

The National Institutes of Health (NIH) and the Centers for Disease Control and Prevention (CDC) help to ensure the safe and ethical conduct of synthetic biology research through promulgation of risk assessment and containment standards for laboratories and investigators. NIH specifically oversees research involving recombinant DNA molecules and receives advice from the NIH Recombinant DNA Advisory Committee (RAC), a group of non-federal experts governed by the openness and public meeting provisions in the Federal Advisory Committee Act.[4] Biosafety standards and requirements of review are set forth in the *NIH Guidelines for Recombinant DNA Research (NIH Guidelines).* The *NIH Guidelines* require risk-based classification and containment for NIH-funded research involving the construction or use of recombinant DNA molecules, as well as organisms and viruses containing these molecules. Synthetic nucleic acids are addressed to the extent that recombinant methods are used in their assembly.[5] NIH is currently considering a proposal to amend the *NIH Guidelines* to specifically include research with synthetic nucleic acids, regardless of whether recombinant techniques are used. NIH published this proposal in March 2009,[6] and, in June 2010, after consideration of public comment, RAC recommended that the NIH Director adopt these changes. CDC and NIH also promulgate a widely accepted industry standard, *Biosafety in Microbiological and Biomedical Laboratories (BMBL)*, which establishes specific procedures for laboratory safety.[7]

CDC, USDA, the Department of Health and Human Services (DHHS), and the Federal Bureau of Investigation (FBI) also play specific roles in addressing concerns about biosecurity. The Federal Select Agent Program (FSAP), administered by CDC and USDA with the Animal and Plant Health Inspection Service (APHIS), regulates individuals and entities possessing, using, or transferring "select agents and toxins" within the United States.[8] Select Agents and toxins are pathogens or biological toxins that have been declared by DHHS and USDA to "have the potential to pose a severe threat to public health and safety."[9] The FBI conducts the security risk assessment of individuals requesting access to Select Agents.[10]

Taken together, these provisions form a protective patchwork quilt of regulations and guidance for research, the workplace, environmental risks, and in some cases pre-market review of safety and efficacy for new products. Anticipated advances in synthetic biology, however, raise questions about the capacity of this system to provide effective oversight of the entire field. Concerns about biosafety and biosecurity, for example, are frequently voiced. Biosafety focuses on protecting people, plants, animals, and the environment from accidental exposure to a pathogen or toxin with potential adverse effects. Biosecurity focuses on keeping biological agents and technologies out of the hands of those who might misuse them. For both biosafety and biosecurity, risk assessment— which typically extrapolates from data on known risks to characterize new and uncertain risks—may be particularly complicated for synthetic biology as novel or previously uncharacterized organisms are developed.

Oversight Challenges

The effective oversight of biotechnology relies on the assessment of the risks posed by the products generated and the process used to generate them. These assessments are predicated on understanding the biologic characteristics of the agent, its host, and the environment in which it will function.[11] In synthetic biology, a major concern is whether the scale of manipulation, using *de novo* chemical synthesis instead of conventional recombinant DNA techniques, raises sufficiently new levels of uncertainty about products, such as their characteristics or safety profile, to warrant new levels or forms of oversight.

The first generation of synthetic biology products is, or may likely be, relatively simple and similar to other genetically engineered products.[12] In the short term, agents generated through synthetic biology are unlikely to raise novel risk assessment or risk management issues. One of the biggest challenges in the oversight of synthetic biology, however, is its capacity to create novel entities that are increasingly dissimilar to known agents or organisms, making potential risks harder to assess. As the field begins to develop more complex, novel, and artificial agents and products, assessing the risks posed will be challenging, particularly for those products with the potential to be released into the environment[13] (see also Chapter 2 for a discussion of risks and benefits). The increasing ease of access to the materials and supplies used to generate synthetic agents poses another unique oversight challenge. Gene and oligonucleotide sequences or parts can be commercially obtained with ease, and reagents and automated equipment for synthesizing nucleic acid sequences are available as well.[14] Deviant uses of synthetic biology could therefore, at least theoretically, occur outside of the scope of existing oversight mechanisms. At this stage, however, technical challenges to creating novel organisms are such that it is difficult to imagine the creation of a substantial threat. Finally, current federal oversight of biotechnology is, in some cases, limited to entities that are owned or funded by the federal government. This means that research currently being conducted using private funds is not subject to some federal oversight.

Federal Authorities

Many federal agencies have jurisdiction over research and production activities involving synthetic biology. The Coordinated Framework organized lead responsibilities for oversight of intentional, beneficial uses of biotechnology, but did not compartmentalize oversight. The oversight is integrated, and overlap is minimized but necessarily exists as the framework is built to respond flexibly to changing science. This shared oversight is described below, generally, in terms of the particular sectors discussed earlier in this report, but the discussion should not be understood to suggest silos or pigeonholes in the oversight system.

Biosecurity

Several regulatory schemes and initiatives are focused on reducing biosecurity risks arising from biotechnology. These include FSAP, export and interstate transfer limitations, and the 2010 guidance for synthetic double-stranded DNA providers.[15]

Federal Select Agent Program

FSAP is administered by DHHS/CDC and USDA's APHIS.[16] Congress established FSAP to limit the possession, use, and transfer of biological agents and toxins, designated as "Select Agents," that have the potential to pose a severe threat to public health and safety, animal and plant health, or to the safety of animal or plant products.[17] Facilities that possess, use, or transfer Select Agents, including for use in synthetic biology, must be registered with FSAP, and individuals or entities seeking to use, transfer, or possess Select Agents must apply for registration and approval for these activities. Select Agents that are regulated by both CDC and APHIS are referred to as "overlap" agents and involve threats to both sectors.[18]

The Select Agent regulations extend both to specific agents as well as certain genetic elements, recombinant nucleic acids, and recombinant organisms. Among the regulated genetic components are: (1) nucleic acids that can produce infectious forms of any of the Select Agent viruses; (2) recombinant nucleic acids that encode for the functional form(s) of Select Agent toxins if the nucleic acids (a) can be expressed in vivo or in vitro, or (b) are in a vector or recombinant host genome and can be expressed in vivo or in vitro; and (3) Select Agents and toxins that have been genetically modified.[19] These regulations are specifically targeted to address scientific advancements such as synthetic biology. The nucleic acid sequence information of Select Agents is not regulated.[20]

Each individual or entity applying for registration must designate a "Responsible Official" who will ensure compliance with the regulations, including conducting annual inspections and overseeing proper disposition of Select Agents.[21] Registration is granted to entities only after a risk assessment is performed for the individuals who have access to, or the ability to gain possession of, a Select Agent or toxin. Additionally, registration is contingent upon a facility inspection by FSAP, and approval of additional documents such as security, biosafety, and incident response plans. Any certificate of registration issued is only valid for three years and for a single physical location, and no individual may access a Select Agent at any time without approval by the DHHS Secretary or the APHIS Administrator.[22] All records relating to Select Agents and toxins must be kept for three years and produced upon request.[23] Any theft, loss, or release of a Select Agent must be reported to the relevant agency immediately and an APHIS/CDC Form 3 must be submitted within seven calendar days.[24] Inspectors may inspect records and premises where Select Agent activities are carried out without prior notification.[25]

APHIS/CDC issued guidance recently for those who create or use synthetic biology and may therefore be subject to the Select Agent regulations.[26] This guidance was partially in response to the National Science Advisory Board for Biosecurity's (NSABB's) report, *Addressing Biosecurity Concerns Related to the Synthesis of Select Agents*, which discusses the regulatory and oversight framework as it relates to synthetic genomics and Select Agents.[27] Elsewhere, the FBI has implemented a "tripwire" initiative in partnership with the U.S. synthetic biology industry to report suspicious requests for genetic sequences. The FBI also has conducted outreach to academia and industry and do-it-yourself (DIY) communities to improve biosecurity for synthetic biology research and uses.[28]

Export Administration Regulations

The Bureau of Industry and Security within DOC administers the Export Administration Regulations.[29] These regulations govern the export and reexport of dual-use commodities, software, and technology from the United States and apply to any individual or entity seeking to export.[30] Included in this group may be researchers collaborating with overseas colleagues, manufacturers with foreign plants, and gene synthesis providers shipping orders outside of the United States. Particularly relevant to the oversight of synthetic biology are provisions designed to restrict access to materials that have dual use applications (i.e., materials with both commercial applications and military or other defense applications).[31]

Items subject to the Bureau of Industry and Security's licensing authority are listed on the Commerce Control List (CCL).[32] Category 1 of the CCL contains "materials, chemicals, 'microorganisms,' and toxins."[33] Products are then classified according to "reasons for control:" (1) national security and dual use; (2) missile technology; (3) nuclear nonproliferation; (4) chemical and biological weapons; and (5) anti-terrorism, crime control, regional stability, short supply, United Nation sanctions, etc.

For each controlled item, detailed licensing requirements and policies for screening potential recipients are imposed. Licenses are provided depending on the nature of the threat. The end user receiving the product must also be screened against lists of proscribed individuals and organizations. Relevant screening lists include: (1) the Entity List (parties who may trigger a license requirement under Export Administration Regulations), (2) the Denied Persons List (parties denied export privileges), (3) the Unverified List (parties where the Bureau of Industry and Security has been unable to identify the end user in prior transactions), (4) the Specially Designated Nationals List (parties barred by the Treasury, Office of Foreign Assets Control), (5) the Debarred List (parties barred by the State Department under International Traffic in Arms Regulations), and (6) Nonproliferation Sanctions (parties that have been sanctioned under various statutes).[34]

Interstate Transfer Regulations

DOT sets rules for the safe and secure transportation of hazardous materials,[35] which may encompass materials necessary for, or created by, synthetic biology. Designated hazardous materials include substances (e.g., wastes and pollutants) that DOT believes are capable of posing an unreasonable risk to health, safety, or property during transport.[36] All persons transporting hazardous waste by air, rail, highway, or water must follow the regulations put out by the Pipeline and Hazardous Materials Safety Administration (PHMSA), an agency within DOT. Specified packaging, labeling, and transport requirements are imposed. For example, packages containing hazardous waste must be able to withstand conditions normally involved in transportation such as changes in pressure, temperature, and humidity, as well as vibrations and shocks.[37] Hazardous materials must also be labeled appropriately to warn transporters (and possible emergency responders) of the type of material contained in the packaging.[38] Hazardous material employees must receive training on PHMSA regulations so they can perform their functions safely.[39]

PHMSA is authorized to conduct inspections and enforce these regulations with civil penalties. Inspectors may send warning letters alerting transporters to probable violations or issue citations if they believe the alleged violation does not have a "direct or substantial impact on safety."[40] Any person who knowingly violates a requirement of these regulations may be liable for a civil penalty up to $55,000 per transportation or shipping violation, and up to $110,000 if the violation results in death, serious illness, serious injury, or substantial destruction of property.[41] In addition, anyone who knowingly, willfully, or recklessly violates the regulations and releases a hazardous material may be imprisoned for up to 10 years for any resulting death or bodily injury.[42] DOT reserves additional authority through the Federal

Motor Carrier Safety Administration, the Federal Railroad Administration, the Federal Aviation Administration, and the U.S. Coast Guard.[43]

Screening Framework Guidance for Providers of Synthetic Double-Stranded DNA

In October 2010, DHHS issued guidance for screening orders of synthetic double-stranded DNA. The guidance addresses the potential biosecurity concerns associated with the use of double-stranded DNA synthesis to reconstruct regulated pathogens and toxins. The guidance recommends "baseline standards...regarding the screening of orders so that they are filled in compliance with current U.S. regulations and to encourage best practices in addressing biosecurity concerns associated with the potential misuse of their products to bypass existing regulatory controls."[44] Compliance with the guidance is voluntary, but many of its specific recommendations reflect underlying statutory or regulatory mandates. Meeting its standards will help ensure that synthetic double-stranded DNA provided for use in synthetic biology will be in compliance with applicable federal regulations, namely the Select Agent regulations and the export administration regulations.[45]

The guidance emerged from a multi-year, public engagement process. Designed as "best practices," the intention of the guidance is efficient update as new information and technical skills emerge. The drafters explained: "[t]he target audience for this guidance is the gene and genome synthesis industry, because the technical hurdles for *de novo* synthesis of Select Agents and Toxins from double-stranded DNA are much lower than for *de novo* synthesis of these agents from single-stranded oligonucleotides."[46] This guidance proposes a screening framework for "commercial providers of synthetic double-stranded DNA that is 200 base pairs...or greater in length to address concerns associated with the potential for misuse of their products."[47] The framework includes "customer screening and sequence screening, follow-up screening as necessary, and consultation with U.S. Government contacts, as needed."[48]

Biosafety

Biosecurity and biosafety concerns frequently overlap, as do the oversight strategies employed to address them. In the earliest days of the genetic engineering era, oversight efforts focused on safety concerns shared by the public as well as scientists conducting this novel research. As the field has grown and matured in the 40 years since then, the tools developed to address these concerns have evolved as well.

NIH Guidelines for Research Involving Recombinant DNA Molecules

NIH established the *NIH Guidelines* in 1976. They were created in light of public concern about emerging techniques for manipulating genetic material, and the 1975 Asilomar Conference on Recombinant DNA in which scientists from academia, industry, and government came together to establish shared principles for containment and safety in such research. The *NIH Guidelines* specify practices for constructing and handling recombinant DNA molecules and organisms and viruses containing recombinant DNA molecules. Compliance with the *NIH Guidelines* is mandatory for investigators at institutions receiving NIH funds for research involving recombinant DNA[49] and would encompass synthetic biology falling within these confines as well. With input from NIH RAC, NIH has modified the *NIH*

Guidelines nearly 30 times since their inception in order to keep pace with advances in science and biosafety. Satisfying their terms is a condition of NIH funding, and they are also widely accepted and followed voluntarily by scientists and organizations, both public and private, across the research enterprise. In addition, other government agencies, including DOE, the Department of Veterans Affairs, and USDA, currently have policies in place that state that all recombinant DNA research conducted or funded by those agencies must comply with the *NIH Guidelines*.[50] Through an active process of public engagement and deliberation, they have become a "gold standard" that is cross-referenced by numerous resources, including *Biosafety in Microbiological and Biomedical Laboratories* (*BMBL*) (see discussion below).[51]

The oversight process prescribed in the *NIH Guidelines* begins at the local level. Through the work of Institutional Biosafety Committees (IBCs)— local groups that include experts in safety and scientific practice—individual research plans are reviewed on a regular basis to assure that biosafety protections, including laboratory containment, are appropriate for the risk posed. Minimum containment measures (Biosafety Levels [BSL] 1-4) based on the known and unknown risks of particular experimental agents and designs are set forth in the *NIH Guidelines*, and institutions may impose additional measures as deemed necessary to comply with their responsibilities under the guidelines.[52] Many IBCs also review other forms of research that entail biosafety risks as part of their institutionally assigned responsibilities.[53] Generally, NIH places primary responsibility on institutions to conduct oversight locally and non-compliance is expected to be self-reported.[54]

Following the advice of NIH RAC and other experts, in 2009 NIH proposed to clarify the scope of the *NIH Guidelines* to specifically cover nucleic acid molecules made solely by synthetic means. The proposed revisions, which are undergoing final review, aim to clarify the applicability of the *NIH Guidelines* to research with synthetic nucleic acids and to provide principles and procedures for risk assessment and management of such research.[55] NIH expects to finalize these amendments this year.

Private work may also interconnect with the federal oversight structure if the institution receives federal research funds. For example, although the work done by the J. Craig Venter Institute on the self-replicating synthetic genome was not federally funded, the Venter Institute is a major federal grant recipient, and thus, is required to adhere to the *NIH Guidelines*, regardless of the source of funding for a particular project.[56] In application, this means that along with IBC review, the Venter Institute followed the corresponding risk group and biosafety measures for the organisms it was working with as prescribed by the *NIH Guidelines*. In addition, the Venter Institute also must follow regulations directed toward private workplaces, such as the OSHA laboratory standards described below.

Biosafety in Microbiological and Biomedical Laboratories (BMBL)

CDC and NIH developed *BMBL* to address the "safe handling and containment of infectious microorganisms and hazardous biological materials,"[57] including those which may be used for synthetic biology. *BMBL* centers on the principles of containment and risk assessment. Containment under *BMBL* includes the "microbiological practices, safety equipment, and facility safeguards" required to protect people who work with biological material, the public, and the environment from exposure. Risk assessment allows the "appropriate selection of microbiological practices, safety equipment, and facility safeguards" required to prevent what *BMBL* deems "laboratory-associated infections."[58] *BMBL* complements the *NIH Guidelines* and is broader in its focus. Laboratories that receive federal funding for research may be required to comply, if, for example, the agency requires compliance as a policy matter for its intramural labs, or as a term and condition of specific extramural funding. There is no federal law that requires compliance for all researchers regardless of funding. Thus, generally, they set a voluntary standard.[59] Biosafety standards evolve as scientific knowledge progresses, and *BMBL*, like the *NIH Guidelines*, is intended to evolve and adapt.

Workplace Oversight

OSHA regulates working conditions for employees in most private sector and federal government workplaces.[60] In addition, many states (State Plan States) have occupational safety and health programs that have been approved by federal OSHA and cover public sector (state and local government) as well as private industry employers.[61] Therefore, the regulations of OSHA or an equivalent State Plan State program will be relevant to most synthetic biology laboratories or workplaces. Under OSHA, employers must create an environment that is "free from recognized hazards that are causing or are likely to cause death or serious physical harm."[62] The regulations of OSHA or an equivalent State Plan State program lay out safety principles and precautions for working with and disposing of hazardous chemicals as well as toxic and hazardous substances. Particular attention is paid to ventilation, sanitation, protective equipment, machinery, and emergency procedures.[63] Hazardous waste cleanup and first aid procedures are also imposed.[64] Employers must evaluate the hazards of chemicals at their work place and inform employees about potential harms through "comprehensive hazard communication programs" including container labeling, warnings, and safety data sheets.[65]

The regulations of OSHA or an equivalent State Plan State program also protect employees who may be exposed to blood or other potentially infectious materials such as human bodily fluids, human unfixed tissues or organs, and Human Immunodeficiency Virus (HIV)-containing cells or culture medium.[66] Regulations require employers exposing employees to such substances to have exposure control plans, delineated methods of compliance, and special protocols pertaining to the HIV and the Hepatitis B virus.[67]

EPA also plays a role in workplace oversight through the Toxic Substances Control Act (TSCA). under TSCA, EPA assesses risks to workers from exposure to new intergeneric microorganisms. EPA can impose personal protective equipment requirements and engineering control restrictions to control worker exposure to potentially harmful substances.[68]

Energy

Oversight provisions for synthetic biology in the energy sector include the general security and safety standards described above. They also include specific provisions aimed at various products, for example, biofuels, biosensors (for various applications), and chemical oil dispersants. These provisions may also apply in other sectors, such as health or agriculture as well.

New Chemicals Including Microorganisms

EPA, under TSCA, regulates new chemicals and microorganisms, including those that could be derived from recombinant DNA technologies and synthetic biology.[69] These new chemicals and new microorganisms can have uses in the energy sector but TSCA also addresses other industrial and commercial applications. under the law, individuals or entities seeking to market or import new chemicals or microorganisms into the united States for commercial purposes must give EPA notice. New microorganisms subject to this requirement include "'intergeneric' microorganisms (including bacteria, fungi, algae, viruses, protozoa, etc.) formed by combining genetic material from organisms in different genera" and "microorganisms formed with synthetic DNA not from the same genus."[70] At least 90 days notice and submission of any known or "reasonably ascertainable" data on the intergeneric microorganism are required.[71] EPA scientists then conduct a risk assessment to ensure that the microorganism will not present an unreasonable risk of injury to health or the environment. EPA reviews the proposed use(s) of the new intergeneric microorganism. It evaluates potential human health and environmental hazards as well as potential environmental, worker, and general population exposures from manufacturing, processing, use, and disposal. EPA may require that additional data be developed by the submitter to enable it to make a reasoned evaluation and may limit or impose restrictions depending on the findings of the risk assessment weighed against the benefits of the microorganism.[72] The same process applies to proposed commercial research and development testing of new microorganisms that are released into the environment. Individuals or entities seeking to conduct such field trials must also file 60 days notice and data with EPA.[73]

Certain intergeneric microorganisms are exempt from the requirement for full notification if the manufacturer meets specified criteria defining eligible microorganisms and specified use conditions (including conditions relating to containment, inactivation, and a number of criteria on the introduced DNA). The limited set of microorganisms eligible for exemption are those that have undergone categorical risk assessment as a species, or as a group of strains within a species, whereby specific features of the category and the history of safe use of members of the exempt category were reviewed. The criteria used, and list of eligible microorganisms, were subject to public comment at the time of proposal and had significant input from major scientific societies. This exemption is most applicable to the use of microorganisms to manufacture specialty and commodity chemicals. Also exempt are intergeneric microorganisms used for documented research in contained structures or research required to comply with the *NIH Guidelines*. The exemption for research and development conducted in contained structures must also address inactivation controls that take into account considerations such as the organism's ability to survive in the environment, potential routes of release, and procedures for transfer of materials between facilities.[74]

EPA's oversight of synthetic biology under TSCA may be limited in ways that pose particular challenges as synthetic biology evolves. First, the amount of information EPA requires to be submitted with a notification that is useful for assessing the risks of microorganisms is limited.[75] Manufacturers need not test new chemicals for toxicity, pathogenicity, or other harmful effects before they submit a notification to EPA.[76] Therefore, EPA may have limited information on which to base its risk assessment. However, if EPA determines that the available information is insufficient to permit a reasoned evaluation of the health and environmental effects of a new intergeneric microorganism and that the microorganism may pose an unreasonable risk, EPA typically will allow submitters to suspend the notification review period to enable such data to be developed. EPA can also impose restrictions on the manufacture, processing, distribution in commerce, use, or disposal of a new intergeneric microorganism to limit exposures and releases until sufficient data are developed. However, with the potential for increasing complexity with synthetic biology products, predictability of the properties of microorganisms will be more complicated. Under TSCA, EPA does require immediate reporting by industry of new information on existing substances which reasonably supports the conclusion that the substance presents a substantial risk to human health or the environment.[77]

A second challenge is that the reach of the law is limited to commercial or commercial research and development activities.[78] It is unclear that all potential users or developers of synthetic biology products, for example, noncommercial research efforts by DIY users, are covered.

Human Health

FDA is the primary regulatory agency that exercises specific authority over drugs and devices for human health. Research activities related to such products are also subject to concurrent biosecurity and biosafety protections described above, including the *NIH Guidelines*, *BMBL*, and TSCA rules of EPA, as applicable.

Food and Drug Administration

New drugs and devices must satisfy FDA's safety and effectiveness standards before they can be introduced into the U.S. market.[79] For drugs, these standards require pre-market review and approval. For devices, FDA requires manufacturers to show substantial equivalence to a marketed device. FDA regulated foods, discussed below, and cosmetics generally reach the market without pre-market approval, although food additives and colorings are reviewed.

Synthetic biology potentially may be used in some fashion in all of the products that FDA regulates. For decades in the health care area, FDA has reviewed and approved numerous biotechnology-derived pharmaceuticals and devices, including drugs and devices created from bioengineered organisms.[80] It approved its first recombinant product, human insulin, in 1982.[81] FDA has issued guidance to explain its thinking about the application of its laws and regulations in the biotechnology sector.[82] It draws no distinction between traditional recombinant techniques and synthetic techniques for genetic engineering. Gene segments "may be obtained from other organisms, or synthesized from scratch in a laboratory."[83]

Before, during, and after approval for clinical testing or marketing, the manufacturer or researcher (e.g., the "sponsor") of a product must work closely with FDA and provide ongoing data about safety and effectiveness.[84] FDA also oversees pre-clinical testing, manufacturing processes, and advertising and promotional labeling.[85] The agency imposes minimum ethical standards on clinical research that it oversees, including requiring the informed consent of research participants, safety and ethics review at the local level, and adverse event reporting.[86]

FDA retains considerable and ongoing authority to monitor safety and protect consumers. It may require manufacturers to submit a "risk evaluation and mitigation strategy," including use of patient registries or screening tests, to manage known or potential risks at the time of approval or after the product has gone to market.[87] It may withdraw approval, urge a voluntary recall, petition a court for injunction or seizure, require label changes, or issue warnings if so warranted.[88] Severe penalties may also be imposed on violators of FDA's requirements. Perpetrators can face civil penalties up to $1,000,000 per violation and criminal sanctions including up to 10 years imprisonment.[89]

As new technologies and applications arise, such as those that may be created by synthetic biology, FDA has responded and clarified its oversight. In 1985, it held that research using recombinant DNA technology should follow safety and containment provisions of the *NIH Guidelines*.[90] Genetically engineered animals, which may be used, for example, to produce pharmaceuticals or food for human and animal consumption, are regulated under FDA's animal drug provisions because the genetically engineered construct (the modified DNA produced by traditional recombinant or synthetic means) itself is an article that meets the definition of a "drug," something "intended to affect the structure or any function of the body of...animals."[91] FDA's new animal drug approval process, similar to the process for human medicines, generally requires pre-market review and approval. For genetically engineered animals of a species not traditionally consumed as food, and for which animal health and environmental risks are shown to be low, FDA may exercise "enforcement discretion" and decline to require pre-market approval, as it did with aquarium fish engineered to glow in the dark.[92]

Agriculture, Food, and Environment

Many of the laws and regulations discussed above apply to research and commercial activities involving synthetic biology in the agriculture, food, and environment sectors. Under TSCA, for example, EPA undertakes prior review of new chemical substances, like biofertilizers, and other environmental applications of biotechnology, like bioremediation and mineral extraction. In addition, algae developed for chemical production other than energy, when grown in the open, would be construed as a potential environmental release and receive TSCA oversight. Major oversight programs involving plant and animal pests and pesticides are administered concurrently by USDA and EPA respectively. FDA oversees certain food safety and production activities.

Environmental Impact

Under the National Environmental Policy Act (NEPA), all federal agencies undertaking major action must take into account the impact their action may have on the environment.[93] Before reaching a final decision on any proposed action that may have a significant effect, the government must evaluate, through a public process, the anticipated environmental impact of the action along with any reasonable alternatives.[94] Public comment is requested at several points in this process.[95] Pursuant to NEPA and the Clean Air Act, EPA reviews all "environmental impact statements," and makes its comments available to the public. EPA also reviews selected environmental assessments.[96] NEPA does not require agencies to select the alternative with the least environmental impact.[97] The NEPA process, however, helps ensure that agencies are making informed decisions, responding to public concern, and taking into account mitigation of environmental impact. Typically used in situations such as new construction or major changes in federal land use, NEPA requirements may also be applied to laboratory research and scientific advancements. While the drafting of an environmental impact statement is time consuming, the NEPA process adds an important layer of protection to uncertain or controversial decisions surrounding synthetic biology.

Plant and Animal Pests

uSDA's APHIS is responsible for regulating the introduction (importation, interstate movement, and environmental release) of genetically engineered organisms that are known to, or could, pose a plant pest risk.[98] Genetically engineered organisms are considered to be "regulated articles" if the donor organism, recipient organism, vector, or vector agent used in their creation is known to be a plant pest, the plant pest status of that organism is not known, or there is a reason to believe that one of these organisms may be a plant pest, and therefore may encapsulate synthetically created organisms.[99] APHIS derives the authority to regulate the introduction of genetically engineered organisms from the Plant Protection Act of 2000.[100] This act defines a plant pest as a living stage of an organism (such as an insect, bacterium, fungus, or virus) that "may directly or indirectly injure, cause damage to, or cause disease in any plant or plant product."[101] The regulations apply to genetically engineered microorganisms, insects, and other traditional types of plant pests and to any genetically engineered plants if plant pest organisms (bacterial and viral plant pathogens) are the donor organisms and vector agents are used in the creation of these genetically engineered plants.[102]

APHIS currently uses a permit and notification system to authorize the introduction of regulated articles; all regulated articles are eligible for the permitting procedure, and certain regulated genetically engineered plants are eligible for the notification procedure.[103] The notification procedure is an administratively streamlined process. Currently, most regulated genetically engineered plants are introduced under notification, and approximately 10 percent of APHIS authorizations are done under the permitting procedure. A permit may be withdrawn where any permit condition established by APHIS is violated.[104]

In making a regulatory determination for a permit or notification for a regulated article, APHIS bases its determination on whether the actions under notification or permit are unlikely to result in the introduction or dissemination of a plant pest. This determination takes into account various risk factors, including, among other things, a low risk that the genetically engineered organism or its progeny can persist, reproduce, or establish without human assistance.

A person may petition the agency to evaluate submitted data and assess whether a particular regulated article is unlikely to pose a plant pest risk, and, therefore, should no longer be subject to APHIS regulations for genetically engineered organisms.[105] If, based on submitted information, the agency concludes that the article is unlikely to pose a plant pest risk, the agency may make a determination to approve the petition and confer non-regulated status on the regulated article. Thereafter, APHIS would no longer require permits or notification for the introduction of this genetically engineered organism.[106]

For animals and genetically engineered animal products, APHIS controls import, export, and interstate movement through a similar licensing process. To apply for a product license, test reports and research data must be submitted that establish the purity, safety, potency, and efficacy of the product. Product labels, including all claims made on them and in advertisements, must also be submitted.[107] Facility licenses are approved once a USDA administrator has approved the conditions of the production facility and production methods and verified that the applicant is sufficiently qualified.[108] Researchers and sponsors must show that their experimental product will not contaminate any current products and will be carefully disposed of and controlled.[109] Authorization to ship experimental products is allowed only in very strict circumstances and to limited destinations.[110]

No products may be imported into the United States without a permit.[111] Biological product permits can be issued for research and evaluation, distribution and sale, and transit shipment.[112] Strict requirements for containment, disease profile of the shipping country, qualifications of the recipient, and safety of the product, among others, are applied during the application process.[113] Ongoing inspection of production facilities and products may be undertaken.[114] Manufacturers and importers must keep detailed records of the production process, testing results, and inventory and disposition of the product.[115] These detailed APHIS regulations would therefore add many helpful pieces to the patchwork quilt of protections for different types and uses of synthetic biology.

Pesticides

Before pesticides can be commercialized or used in the United States, they must meet specific health and safety standards under the Federal Insecticide, Fungicide, and Rodenticide Act (FIFRA).[116] FIFRA requires EPA to determine that a pesticide will not pose an unreasonable risk of harm to human health or the environment.[117] Both naturally occurring and genetically engineered microorganisms and plants, including those created by synthetic biology, are regulated in this way.

EPA's pre-market approval and post-market adverse event reporting requirements rely on careful scientific evaluation. The agency must determine "with a reasonable certainty" that "no harm to human health" and no "unreasonable risks to the environment" will occur when the product is used as intended and according to label directions.[118] EPA requires applicants to perform various tests and submit comprehensive data before approval.[119] EPA also sets "tolerances," meaning maximum pesticide residue levels, for the amount of the pesticide that can remain in or on foods or feed crops.[120]

For research on pesticides, EPA's oversight is more limited. It does not require pre-approval for laboratory or contained-setting research.[121] Field testing, which is a prerequisite for commercial marketing approval, usually requires EPA pre-clearance through an experimental use permit or notification.[122] Experimental use permits are granted if, in EPA's view, the experimental use will not yield unreasonable adverse effects on the environment. As with marketing approval, applicants must submit detailed information including safety and pre-field testing data, to support their permit request.

States also regulate pesticides under FIFRA and applicable state laws. Some states impose more restrictive requirements and others defer to EPA's oversight.[123]

Genetically Engineered Foods

All genetically engineered animals, regardless of whether they are intended for food use, are within FDA's jurisdiction, as explained above, because the recombinant DNA constructs that alter the animal's structure or function meet the definition of new animal drugs. FDA oversees the safety and effectiveness of these animals through its pre-market review and approval processes. Applicable law does not require pre-market clearance for "food," whether derived from plant or animal. However, FDA requires evidence that food additives are safe at their intended level of use before they may be added, which is relevant for the products of genetic engineering.[124]

FDA has two main authorities over foods. First, it has post-market authority to seize foods that pose a risk to public health.[125] Second, it may regulate as food additives the substances (e.g., enzymes) added to plants. For example, in 1994 the agency reviewed a genetically engineered tomato with improved ripening qualities and regulated a gene product added to the tomato as a food additive.[126] Where a substance is not "generally recognized as safe" or otherwise exempt, FDA must review and approve the use of the additive before marketing, regardless of the technique used to add it to food.[127]

FDA is authorized to assure that the foods under its purview bear labels that are truthful and not misleading.[128] For foods from genetically engineered plants, FDA policy expressly indicates that name changes are appropriate only if "the resulting GE [genetically engineered] food product" is "materially different from its traditional counterpart," meaning that "the GE food product differs in nutritional quality, taste, etc."[129] In the tomato example cited above, FDA found use of the traditional name "tomato" appropriate because the genetically engineered product did not meaningfully differ in chemical composition from a traditional tomato. In contrast, FDA did require a special label for oil derived from a genetically engineered soybean plant because it contained significantly higher amounts of oleic acid than traditional soybean oil.[130] Production methodology (i.e., whether a product is produced through biotechnology or through conventional breeding) is not considered "material" information, and therefore such information is not required to be disclosed on the food label.[131] FDA follows this same standard for foods from genetically engineered animals, although no genetically engineered animals have been approved for food at this time.

Environmental Impact and Clean up

At the far end of the oversight scheme, particularly at this early stage of synthetic biology research and development, are remediation programs. EPA oversees programs for prevention and emergency management of chemical accidents;[132] oil pollution prevention and discharge;[133] and emergency planning and notification.[134] Under EPA, the Office of Solid Waste and Emergency Response also has both emergency and long-term clean up programs under the Comprehensive Environmental Response, Compensation, and Liability Act, as well as the Resource Conservation and Recovery Act. The scientific risk assessment and response strategies employed in these operations are likely to evolve as the field of synthetic biology itself evolves.

Summary

Multiple federal departments and agencies have significant oversight responsibilities for synthetic biology. The scope of these authorities extends from the laboratory to the field, the environment, the workplace, and the market. Some agencies impose specific safety conditions on research funded with federal dollars or at institutions that receive federal funds. Others reach all research, development, and commercial activities that raise specific threats or risks of harm. Generally, there is at least one federal agency—NIH, CDC, FDA, USDA, OSHA, DOT, DOC, or EPA—with specific oversight responsibility for a proposed application and frequently there is overlapping jurisdiction. Where prior experience or the character of the activity warrants heightened scrutiny, like drug and device development or pesticide use, pre-market review or approval is usually required. Genetically engineered animals require approval by FDA prior to entering into commerce.

This patchwork quilt of measures is built on long-standing practices that have adapted to new technologies over time. Risk assessment in this field may be particularly challenging and require both new techniques and new standards. Further adaptation and restructuring may be required as the applications of synthetic biology grow and their consequences are better understood. As elaborated on in Chapter 5, the Commission's overview has indicated that the government should undertake a more comprehensive review, through a central body such as the Executive Office of the President, to assure that the existing patchwork quilt is indeed affording the u.S. public and the environment with adequate protections as the field of synthetic biology advances.

Analysis and Recommendations

The President asked the Commission to recommend how the developing field of synthetic biology and related technologies can maximize public benefits, minimize risks, and observe appropriate ethical boundaries. A framework of basic ethical principles can provide guidance in the assessment of an emerging technology such as synthetic biology. In this case, as described in Chapter 1, five principles are identified that are most relevant to assessing ethical considerations related to synthetic biology and other emerging technologies:

1) Public Beneficence
2) Responsible Stewardship
3) Intellectual Freedom and Responsibility
4) Democratic Deliberation
5) Justice and Fairness

The Commission relied on these principles to conduct its analyses and build its recommendations, as presented in this chapter. It is the Commission's hope that these principles will be applicable not only to synthetic biology, but also to assessing other emerging technologies.

Public Beneficence

The ideal of public beneficence is to act to maximize public benefits and minimize public harm. This principle encompasses the duty of a society and its government to promote individual activities and institutional practices, including scientific and biomedical research, that have great potential to improve the public's well-being. In the case of emerging technologies like synthetic biology, this improvement may be by means of providing improved or more widely available forms of medical and health care, food, shelter, transportation, clothing, and eco-friendly fuel, along with other means of improving people's lives. Scientific and technological discovery often have the added potential of increasing economic opportunities, which also redound to the public good.

This section focuses on how society and its members—individually and collectively—can provide an environment for synthetic biology to flourish for the benefit of as many people and communities as possible. The Commission observed during its deliberations considerable enthusiasm for the field among scientists, industry representatives, and the public. The anticipated benefits portend dramatic potential improvements in energy production, the economy, health care, and other areas that would enhance public welfare. The development of strategies that will allow the field to continue to grow in ways that offer the greatest potential net benefit to individuals and communities, both in the United States and worldwide, should be a high priority for public policy.

Promoting Public Well-Being and Prioritizing the Public Good

Citizens and their representatives have good reason to be engaged observers in the development of synthetic biology, particularly in light of the potentially transformative benefits to society of potential uses. Chapter 3 presented current examples of synthetic biology applied in research and development programs designed to benefit humankind. Environmentally friendly biofuels and affordable antimalarial drugs are among the near-term products of synthetic biology already receiving significant attention. These are important current examples of how advances in synthetic biology may deliver widespread benefits that promote social welfare. Continued investment in this field should be directed to these types of applications and others that offer similarly expansive opportunities to address serious problems that affect our collective well-being.

The Commission's deliberations called attention to the diversity of interests and practitioners participating in the synthetic biology community. Despite their range of disciplinary backgrounds, nationalities, and institutions, synthetic biologists appear united in contributing their expertise to the development of novel products that address global needs. Distinguishing between academic, public, and commercial research in synthetic biology is extremely difficult, as many researchers are active contributors in each domain. In many ways, drawing this distinction is unnecessary. The organizational home of an individual practitioner may not limit his or her ability to work with others to accomplish shared research goals.

This intermingling of academic and commercial research—both basic and applied—provides fertile ground for innovation.[1] The development of semi-synthetic artemisinin, an antimalarial drug, is one example that demonstrates how academic, public, non-profit, and industry interests have come together to promote global well-being. In this case, researchers at a public university interested in exploring synthetic biology identified the production of artemisinin, a treatment for malaria, as potentially improvable using synthetic biology techniques. An estimated one million people, primarily children under the age of 5 years, die annually from malaria.[2] Researchers began with public dollars and expanded their work in partnership with a private foundation. The results are being commercialized by a for-profit pharmaceutical manufacturer, and a non-profit foundation is planning for eventual distribution. While the story is not over and initial drug production remains in process, the model shows how collaboration between academia, the private sector, and industry can use synthetic biology to address significant societal problems.

The artemisinin story illustrates one way that a diverse group of interests and funding sources—both public and private—can collaborate on research and development activities involving synthetic biology. As with many emerging technologies at an early stage, however, public information about the amount of public and private investment in this field is minimal.[3]

Public funding of research can bring an enhanced measure of focus, oversight, and accountability to any emerging technology. Absent national security protections, government-funded research in the United States is publicly disclosed. Public funding also promotes transparency and accountability that might not exist in purely private efforts.

Private funds may not be widely available for research into risk assessment practices or the ethical and social safeguards that aim to maximize public benefit while minimizing the risks of new technologies like synthetic biology. In synthetic biology, there are some notable exceptions of private funding for efforts to examine ethical, legal, and social issues, including commendable activities supported directly by the J. Craig Venter Institute and the Alfred P. Sloan Foundation.[4] The scope and impact of these efforts are, however, generally quite limited. In order to understand the possibilities and moral limits of synthetic biology public funding may be necessary to augment such efforts. The products of this work are critical to ongoing efforts to evaluate safety and to promote public acceptance of this emerging field. Research exploring the normative and conceptual issues related to these topics can be a valuable complement to the empirical, quantitative, or qualitative work that typically receives greater support from public funding sources.

To promote public engagement and assure needed transparency regarding federal efforts in the field of synthetic biology, the government should review and make public findings regarding the scope of its research funding at this time.

Recommendation 1: Public Funding Review and Disclosure

Through a central body such as the Executive Office of the President, the federal government should undertake a coordinated evaluation of current public funding for synthetic biology activities, including funding for research on techniques for risk assessment and risk reduction, and for the study of ethical and social issues raised by synthetic biology. This review should be completed within 18 months and the results made public.

The evaluation recommended here would ensure effective use of public funds, promote transparency, develop priorities, and avoid redundancy. This recommendation and the examples below align with the Commission's interest in justice and fairness. It aims for the potential benefits of synthetic biology to extend to as many individuals and communities as is reasonably possible, addressing this country and the world's most urgent and compelling needs. Public funding can be an important tool in realizing these goals and doing so in ways that are sensitive to ethical and safety concerns.

Most potential products of synthetic biology are in very early stages of development. Basic research is critical to further expansion of this science and its effective translation into useful products. Basic research—work that focuses on enhancing our understanding of fundamental principles of science and the natural world—is also important to the growth of the field. Direct commercial applications are not typically the intended outcomes of basic research, yet this work can also be extremely valuable to society. A commitment to basic research reflects a belief that knowledge is itself a public good.

More practically, scientific fields invariably develop in unanticipated ways. The aggressive pursuit of fundamental research generally results in a broader understanding of a maturing scientific field like synthetic biology than approaches solely focused on developing specific applications to address contemporary needs. This understanding of basic principles may be a particularly valuable way to prepare for the emergence of unanticipated risks that would require rapid identification and creative responses.

At the same time, synthetic biology research is in competition for scarce resources with other areas of science and other societal needs. Decisions will be required regarding which research directions deserve funding over others. These decisions should be driven in part by which strategies offer the most promise based on scientific, technical, and social considerations.

Potential profitability is also a significant motivator of research and development investments. When research is fairly new, as in the emerging field of synthetic biology, the promise is often high but the incentives for investment can be low because of uncertain success or marketability. Some drugs that address asymptomatic risk factors or "lifestyle" issues (e.g., drugs that do not treat life threatening conditions or pain), rather than specific disease processes, have received significant attention from the pharmaceutical industry because there is a large market and potential for profit in the United States and other developed nations. Some of these drugs can be quite beneficial to patients, such as statins to reduce elevated cholesterol levels. Drug manufacturers frequently devote research resources to the development of very similar versions of competitors' already successful, profitable products instead of pursuing novel research directions with a less certain path to success or profits.

Other more prevalent and deadly diseases lacking therapeutic or preventive options today receive lower investment priority, especially diseases more common in developing nations. Absent a reliable market in the United States or other wealthy countries, manufacturers often choose not to devote significant investment dollars to these diseases, choices reflecting rational responses to the market. Government and others interested in promoting public well-being, such as private foundations, can effect change by re-drawing the financial landscape for research and development in these areas.

Recent congressional and Administration emphasis on "high risk/high reward" research offers one example of how the public good can be promoted when market forces alone may not succeed. The National Institutes of Health (NIH) has created several programs that specifically support creative, highly innovative research approaches that might otherwise be too novel or too risky to receive funding through traditional channels.[5] In 2007, Congress also expressly directed the agency to award research grants for these types of potentially high-impact research projects.[6] In the private sector, the Bill and Melinda Gates Foundation, through its "Grand Challenges in Global Health" program, is changing the financial picture by awarding substantial grants— nearly $500 million dollars in recent years—to stimulate scientific innovation among traditional and nontraditional researchers to treat and prevent diseases most prevalent in developing nations.[7]

The development of novel antibiotics is one example in which incentives could help stimulate research interest toward an important public need that might otherwise not receive sufficient attention.[8] Similarly, funding or incentives to spur research into age-related degenerative diseases of the nervous system (including dementia and gait disorders) may help in the quest for cures for Alzheimer's disease, Parkinson's disease, and related disorders prevalent in aging populations.

The Commission's deliberations focused specifically on synthetic biology, following its charge from President Obama, but alternative research strategies are also appropriately being pursued to address many of the national and global concerns for which synthetic biology may provide solutions. The Commission supports public and private investment in synthetic biology-related research as one important avenue of research among others.

Recommendation 2: Support for Promising Research

Advancing the public good should be the primary determinant of relative public investment in synthetic biology versus other scientific activities. The National Institutes of Health, the Department of Energy, and other federal agencies should continue to evaluate research proposals through peer-review mechanisms and other deliberative processes created to ensure that the most promising scientific research is conducted on behalf of the public.

Synthetic biology is advancing rapidly. Future funding decisions should be made through ongoing evaluation of the state of the science and its potential applications. Policy makers, the scientific community, and the public should continue to assess the adequacy of existing peer review and funding mechanisms to address future advances in synthetic biology-related science and technology. Private interests, including for-profit and nonprofit entities, should likewise consider global public needs that can be uniquely advanced through their efforts.

Realizing Economic Opportunities

Most current attention to the potential benefits of synthetic biology focuses on applications related to health, energy, and the environment. Investment in synthetic biology also can bring economic benefits, both from the direct activities related to research and development and from the eventual commercialization of successful technologies. These benefits have the potential to strengthen communities through the creation of jobs and other opportunities, thereby enhancing citizens' quality of life. Forecasting the potential impact of synthetic biology on job creation and economic growth is difficult, but the Commission received public comments estimating that the use of synthetic biology in the chemical industry alone could generate global revenue of $1 trillion and create 1.2 million direct jobs.[9] Additional revenue and jobs would be expected from synthetic biology activities related to pharmaceutical and agricultural applications.

Potential economic benefits may be particularly valuable to communities in developing nations, where health, access to resources, and economic stability are closely linked to one another and to disparities in health and welfare. This underscores the importance of adopting a global perspective when considering the potential benefits of synthetic biology.

Although the potential economic benefits cannot be known with precision, this potential should nonetheless be continually assessed as part of activities to promote synthetic biology. Technological solutions alone cannot eliminate the fundamental causes of global inequality, but they can contribute to comprehensive programs to address them. This theme is addressed when considering the principle of justice and fairness.

Intellectual Property and the Sharing of Scientific Knowledge

Information sharing and reasonable access to discoveries and inventions have long fueled the scientific enterprise. These activities enable scientists to leverage each other's work in order to more quickly advance new projects and translate basic research into products. Impediments to innovation and information sharing, some say, arise from the patent and copyright system. These mechanisms afford inventors and authors a time-limited right to prohibit others from using their work or similar work of the same design. One concern consistently raised with regard to biotechnology is the potential limiting effects of intellectual property claims over research results, particularly in basic research.[10] Patents on discoveries and restrictive or exclusive licensing agreements may encourage, but also may deter or increase the development costs of subsequent inventions that build on the basic discovery. Some who provided testimony to the Commission argued that the current system unduly limits scientific advances; others took the opposite view and asserted that the current system works well.

The patent system is designed to encourage innovation and investment by providing incentives to inventors to disclose their discoveries to the public so that others can build on them. In return, the inventor is granted exclusive rights to the invention and control of its development for a limited period of time. Balancing the interests of the inventor and those who wish to use the invention is a challenging task in science generally, and in biotechnology particularly. These concerns have been the focus of numerous studies, for example, in genomics.[11] Ongoing discussions have focused on the roles and responsibilities of government, the academic community, and the private sector in adopting intellectual property practices that foster an environment in which invention and innovation can thrive. Such discussions are likely to continue as patent law and court decisions in this area evolve.[12]

Synthetic biology raises challenging issues in this area as a result of research interest in creating standard biological "parts" that can be combined to build new biological systems or organisms for potential use in health care, agriculture, and energy (see Chapter 3). The field also is particularly dependent on information technology and the need for common standards.[13]

Concerns about the effects of patenting on synthetic biology mirror those expressed about patents involving DNA and genetic tests—that is, whether patents will be granted that are either too narrow or too broad.[14] Overly broad patents could "restrict collaboration and stifle development in the field, and narrow patents may overcomplicate the process, meaning that hundreds of patents have to be negotiated to produce a system from standardized parts."[15] For example, the Venter Institute is seeking a patent on the synthetic cell it described in May 2010 and on processes for making synthetic genomes. For some, these efforts raise questions about the extent to which a patent on synthetic organisms should be issued and whether doing so is in the public interest.[16] Others in the synthetic biology community have taken steps to keep some portion of the "parts" developed with synthetic biology available in an open-source system (e.g., BioBricks and the Registry of Standard Biological Parts) without traditional patent restraints.[17]

In the last 20 years, we have seen increased emphasis on transparency, data sharing, and creative licensing practices for patentable subject matter.

This trend applies especially, though not exclusively, to publicly funded research. Examples of current data sharing requirements include several NIH policies introduced since the late 1990s, most recently the 2007 NIH Public Access Policy, a congressionally mandated provision for public distribution of research results.[18] Similar policies apply to awardees of the Howard Hughes Medical Institute and the Wellcome Trust, private research funding sources in the United States and the United Kingdom, respectively.[19] Public clinical trial disclosure requirements have arisen from the private sector, through research journal publishers, and the public sector, through congressional actions in 1998 and 2007.[20] Demands for licensing inventions to meet social needs, including providing access to medications or enabling more research and technology development, have fueled innovative licensing practices and creative solutions to so-called "patent thickets" and other limitations on scientific exploration.[21]

The principle of public beneficence requires researchers, inventors, patent holders, and others to work together to develop creative strategies to maximize opportunities for innovation.

Licensing alternatives could include methods of compulsory or bundled licensing, patent pooling, and broad, non-exclusive licenses for foundational technology. Because synthetic biology is in large part based on the application of engineering principles through the use of standardized, modular parts, access to those standard components could be especially critical to the development of the field.

Intellectual property issues in synthetic biology are evolving. The Commission offers no specific opinion on the effectiveness of current intellectual property practices and policies in synthetic biology. It recognizes that there are important concerns that deserve ongoing attention, especially as this rapidly developing field evolves. Current litigation, such as *Association for Molecular Pathology, et al. v. U.S. Patent and Trademark Office, et al.*, is likely to influence practices and policies in the future. This case presents the question of whether isolated human genes—those with mutations associated with an increased risk of breast cancer—and the comparison of their sequences is patentable.[22]

Thus, the government should keep careful watch on this field and consider best practices and other policy guidance, if needed, to ensure that access to basic research results and tasks is not unduly limited.

Recommendation 3: Innovation Through Sharing

Synthetic biology is at a very early stage of development, and innovation should be encouraged. The Executive Office of the President, as part of the coordinated approach urged in Recommendation 4, should lead an effort to determine whether current research licensing and sharing practices are sufficient to ensure that basic research results involving synthetic biology are available to promote innovation, and, if not, whether additional policies or best practices are needed. This review should be undertaken with input from the National Institutes of Health, other agencies funding synthetic biology research, such as the Department of Energy and the National Aeronautics and Space Administration, the U.S. Patent and Trademark Office, industry, academia, and public civil society groups. The review should be completed within 18 months and the results made public.

The Commission urges the government to consider subsequent reviews and coordinated assessment if needed. Information sharing is a critical mechanism for promoting scientific progress and innovation.

Responsible Stewardship

The principle of responsible stewardship calls for prudent vigilance, establishing processes for assessing likely benefits along with assessing safety and security risks both before and after projects are undertaken. A responsible process will continue to assess safety and security as technologies develop and diffuse into public and private sectors. Prudent vigilance does not demand extreme aversion to all risks. Not all safety and security questions can be definitively answered before research begins, but prudent vigilance does call for ongoing evaluation of risks of harm along with benefits. The duty to be responsible stewards of nature, the earth's bounty, and the world's safety rests on concern not only for human health and well-being today but also and importantly for future generations and the environment looking forward.

The principle of responsible stewardship can be interpreted in an operational way to pose the question, "What can and should we, as a society, do in response to the emerging field of synthetic biology to be responsible stewards of nature, the earth's bounty, human health and well-being, and the world's safety, now and into the future?"

Options for action in this area range from doing nothing—that is, allowing the field of synthetic biology to proceed without limits or regard for public or environmental safety—to halting or substantially slowing its progress until risks can be identified and mitigated. One common interpretation of the "precautionary principle" would prescribe the latter approach. There are several definitions of the precautionary principle, but it generally states that if an action or policy has the potential to cause harm but uncertainty exists regarding the likelihood or severity of harm, the responsibility for demonstrating the safety of the approach belongs to those advocating for the policy or action.

The precautionary principle evolved primarily in the context of European debates and resolutions concerning the environment and genetically modified foods, and it is often raised in discussions involving risk and uncertainty in public policy in the United States and internationally.[23] One premise behind the precautionary principle may be that because there is a social responsibility to protect the public or the environment from plausible and avoidable harms, protections should be relaxed only when science produces evidence that harm is unlikely to result. In some legal systems, such as that of the European union, the application of the precautionary principle is a statutory requirement.[24]

A contrasting perspective is the "proactionary" principle, which assumes that an emerging biotechnology should be considered "safe, economically desirable and intrinsically good unless and until shown to be otherwise, which means that the burden of proof is on those who want to slow down a given line of research."[25] Advocates of the proactionary principle appeal to a commitment to intellectual freedom, the autonomy of individual decision making, economic growth, national competitiveness, and improved health and well-being. At its most extreme, this principle might allow science and technology to go forward unfettered, but, in general, proponents of this principle have supported some measure of oversight and monitoring.[26]

In order to provide benefits to human conditions and the environment, the Commission thinks it imprudent either to declare a moratorium on synthetic biology until all risks can be determined and mitigated, or to simply "let science rip," regardless of the likely risks. The field of synthetic biology can proceed responsibly by embracing neither the precautionary principle nor the proactionary principle. The Commission instead proposes a middle ground— an ongoing system of prudent vigilance that carefully monitors, identifies, and mitigates potential and realized harms over time. It came to this position for several reasons.

First, synthetic biology does not necessarily raise radically new concerns or risks compared to those that have been expressed about other emerging technologies, for example, molecular biology and nanotechnology. In many ways, synthetic biology is an extension of genetic engineering and part of an increasingly interconnected network of scientific disciplines including, among others, nanotechnology and information technology.[27]

Second, many existing oversight mechanisms and bodies (statutory, regulatory, and voluntary) are well situated and in the process of reviewing and monitoring the field of synthetic biology as it develops. The Commission endorses activities aimed at ensuring that those mechanisms and bodies are sufficiently well coordinated and supported to effectively monitor risks in an ongoing and proactive fashion.

However, synthetic biology does introduce some possible risks that warrant special attention. According to the National Science Advisory Board for Biosecurity (NSABB), synthetic biology poses "varying degrees of uncertainty regarding the predictability of biological properties of partially or completely synthetic agents or materials."[28] It also poses some unusual potential risks, as "amateur" or "do-it-yourself" (DIY) scientists and others outside of traditional research environments explore the field. These risks must be identified and anticipated—as they are for other emerging technologies—with systems and policies to assess and respond to them while supporting work toward potential benefits. In this section, the Commission considers several approaches to promoting responsible stewardship, including oversight mechanisms, establishing safeguards, supporting relevant research, and encouraging and developing a culture of responsibility.

Stewardship through Oversight

Scientists have been conducting biological research that poses risks throughout the history of modern science. Consider Edward Jenner's experiments 200 years ago to develop a smallpox vaccine using cowpox virus, or more recently, gene therapy for rare diseases and studies of pathogens that could kill or sicken thousands through a natural or malevolent environmental release. History tells us that such research has resulted in enormous benefits for society, but it sometimes has had terrible consequences. Over time, safety and security practices and procedures have expanded and evolved to increase the likelihood that risks will be anticipated, mitigated, and monitored and that responses can be activated quickly should harms arise.

In the United States, oversight frameworks already exist for many activities of modern biological science including research involving humans, animals, microorganisms and toxins, and recombinant DNA. Oversight also occurs with regard to laboratory worker safety, use of federal funds in research, and transport and containment of dangerous agents. Oversight is frequently, but not exclusively, tied to public funding or the need to gain regulatory approval in order to market or distribute a product (see Chapter 4).

Long-standing regulatory systems, for example for food, drugs, and chemicals, undergird such approaches, while others developed specifically around the fields of genetic engineering and biotechnology. Some grew out of what were initially, and in some cases remain, voluntary self-policing efforts. These policies tend to be predicated on a risk-benefit assessment that is scaled according to identified risk and that evolves through an ongoing process of open public dialogue.[29] Over time, reflecting principled flexibility, many have been modified as risks, or the lack thereof, became clearer.

Demonstrating the government's increasing attention to this new field, the evolving federal oversight framework for synthetic biology, in the past year alone, includes:

- a proposed revision of the NIH Guidelines for Research Involving Recombinant DNA Molecules to address synthetic biology,[30]
- development of a U.S. government Screening Framework Guidance for Providers of Synthetic Double-Stranded DNA,[31]
- development of an Animal and Plant Health Inspection Service (APHIS)/ Centers for Disease Control and Prevention (CDC) guidance on how current Select Agent regulations apply to those who create and use synthetic genomic products,[32] and
- consideration by the NSABB of strategies for conducting outreach to all practitioners of synthetic biology, enhancing the culture of responsibility, and promoting international engagement.[33]

These efforts build on the existing oversight responsibilities exercised by various federal agencies, including the Environmental Protection Agency (EPA) (chemical safety), the Food and Drug Administration (FDA) (food, drugs, and medical devices), the Department of Agriculture (crops and animal feed), and the Department of Homeland Security (biosecurity).

Internationally, the community of scientists working in synthetic biology, as well as policymakers and ethicists, are also focusing on ways to assure responsible stewardship. For example, the European Commission supports SYNBIOSAFE, a collaborative project among public and private parties that is researching the safety and ethics of synthetic biology. Governance and oversight strategies for synthetic biology research and products are similarly being addressed through multiple efforts at the international level.[34]

To assure responsible stewardship in the field of synthetic biology, clarity, coordination, and accountability must exist across the government. The Commission does not believe that new agencies, offices, or authorities must be developed at this time, if ever. Instead, the Executive Office of the President (EOP) should lead an interagency process to identify and clarify, if needed, existing oversight authorities and to ensure that the government is fully informed on an ongoing basis of developments, risks, and opportunities as this field grows.

Recommendation 4: Coordinated Approach to Synthetic Biology

The Commission sees no need at this time to create additional agencies or oversight bodies focused specifically on synthetic biology. Rather, the Commission urges the Executive Office of the President, in consultation with relevant federal agencies, to develop a clear, defined, and coordinated approach to synthetic biology research and development across the government. A mechanism or body should be identified to: (1) leverage existing resources by providing ongoing and coordinated review of developments in synthetic biology, (2) ensure that regulatory requirements are consistent and non-contradictory, and (3) periodically and on a timely basis inform the public of its findings. Additional activities for this coordinating body or process are described in other recommendations.

These activities might be carried out, for example, under the auspices of the Office of Science and Technology Policy in the EOP, or the Emerging Technologies Interagency Policy Coordination Committee. It is essential that they be coordinated by an office with sufficient authority to bring together all parts of the government with a stake in synthetic biology. It is similarly important that this effort be sufficiently authoritative to effectively engage with, or supervise engagement with, foreign governments. A critical component of this coordinated strategy is to assure both the scientific community and the public that biosafety, biosecurity, and environmental risks of synthetic biology are fully addressed.

In any scientific inquiry, risks must be justified by anticipated benefits. Such balancing of risks and potential benefits is often complicated by uncertainty. Because much of science explores the unknown, policy makers should develop policies that acknowledge uncertainty about both risks and potential benefits. Information, flexibility, and judgment are critical to find the appropriate balance and determine the most responsible way to proceed. The rapid development of the field of synthetic biology makes the challenges of decision making under conditions of uncertainty particularly acute.

Recommendation 5: Risk Assessment Review and Field Release Gap Analysis

Because of the difficulty of risk analysis in the face of uncertainty—particularly for low-probability, potentially high-impact events in an emerging field—ongoing assessments will be needed as the field progresses. Regulatory processes should be evaluated and updated, as needed, to ensure that regulators have adequate information. As part of the coordinated approach urged in Recommendation 4, the Executive Office of the President should convene an interagency process to discuss risk assessment activities, including reasons for differences and strategies for greater harmonization across the government. It should also identify any gaps in current risk assessment practices related to field release of synthetic organisms. These reviews should be completed within 18 months and the results made public.

Individual scientists were among the first to raise concerns about the possible risks posed by synthetic biology research. In fact, synthetic biologists have been discussing among themselves the appropriate safety policies for their fi eld since at least 2004. Members of the synthetic biology community have met in a series of meetings over the past 6 years to discuss concerns about both biosafety and biosecurity. They also considered environmental concerns and appropriate tools for risk assessment.[35] Like SYNBIOSAFE in Europe, the Synthetic Biology Engineering Research Center, a collaborative project funded by the National Science Foundation in the United States, is examining safety, security, and preparedness issues.[36] The willingness and initiative of the scientific community to engage in this level of introspection is both reassuring and essential.[37]

Similar to researchers in the early years of recombinant DNA research in the mid-1970s, those closest to this emerging field have exercised caution. While self-governance is not a sufficient means to mitigate all risks, it is likely an effective way to control many of the risks associated with emerging technologies, including synthetic biology, particularly at this early stage.[38] Individual scientists and students typically are the first to notice the laboratory door ajar, the suspicious behavior, or the lack of safety precautions among colleagues.

The activities of nontraditional scientists involving synthetic biology are also noteworthy. Communities of "amateur" scientists are actively working to increase understanding of potential physical and environmental risks posed by synthetic biology activities. As these communities grow, organized efforts to engage this community in discussions of safety and security and to foster a commitment to responsible stewardship will be increasingly important.

Industry, too, has worked collaboratively to enact policies to promote responsible stewardship. For example, both the International Gene Synthesis Consortium and the International Association of Synthetic Biology—whose members include the vast majority of the gene synthesis providers in the United States and worldwide—have developed best practice guidelines for screening orders and customers. These groups are participating actively in public discussions of regulatory options, collaborating on implementing screening practices, and interacting with the Federal Bureau of Investigation (FBI) on training and notification efforts.[39] Moreover, these organizations and their member companies have committed publicly to improve screening protocols and tools and to incorporate recent U.S. government guidance into practice.[40]

Stewardship through Use of Safety Features and Reviews

Coordination and careful risk analysis are essential steps for responsible stewardship, but they are not sufficient. There are several additional approaches, known today and evolving as our abilities in this field grow, to limit uncertain risks in synthetic biology. Technology can be harnessed to build in safeguards, just as cars have brakes and seatbelts, houses have smoke detectors, and computers have anti-virus software. A number of safety features can be incorporated into synthetic organisms to control their spread and life span.

The intentional and unintentional consequences of novel research designs and new products cannot always be predicted. In the case of a newly engineered synthetic organism, for example, lack of history regarding the behavior of the entity, either environmentally or ecologically, requires that there be a means to track or contain it if it can survive outside of the laboratory.

Surveillance or containment of synthetic organisms is a concrete way to embrace responsible stewardship. These safety features may require some combination of public investment and incentives for additional private funding, and they should be implemented only after they undergo rigorous testing and validation.[41] Promoting and supporting efforts to design and employ safeguards will ensure that they are widely adopted and become a standard tool for practitioners of synthetic biology.

As part of the coordinated approach described in Recommendation 4, and on an ongoing basis as the field progresses, the government should specifically monitor the potential risks of organisms with novel synthetic traits or properties surviving or multiplying in the natural environment. As needed, reliable containment and control mechanisms should be identified and required. Among current options, "suicide genes" or other types of self-destruction triggers could be considered in order to limit the life spans of synthetic organisms.[42] Organisms could also be designed to require nutritional components absent outside the laboratory, such as novel amino acids, thereby controlling them in the event of release. These are options only and should be updated as science progresses. The primary consideration is to ensure that concrete protections are inserted into synthetic organisms to assure safety.

Recommendation 6: Monitoring, Containment, and Control

At this early stage of development, the potential for harm through the inadvertent environmental release of organisms or other bioactive materials produced by synthetic biology requires safeguards and monitoring. As part of the coordinated approach urged in Recommendation 4, the Executive Office of the President should direct an ongoing review of the ability of synthetic organisms to multiply in the natural environment and identify, as needed, reliable containment and control mechanisms. For example, "suicide genes" or other types of self-destruction triggers could be considered in order to place a limit on their life spans. Alternatively, engineered organisms could be made to depend on nutritional components absent outside the laboratory, such as novel amino acids, and thereby controlled in the event of release.

The timing of deliberate release of synthesized organisms into the environment and the need to analyze risks prior to release raises special concern. We must proceed carefully, particularly when the probability or magnitude of risks are high or highly uncertain, because biological organisms may evolve or change after release.[43] Generally, the paradigm for risk assessment throughout the scientific community and oversight agencies is to evaluate a new organism in terms of known relatives and to set containment rules or environmental risk mitigation strategies based on the applicable rules for the known relative (see p. 83). This approach appears to have worked effectively and enabled risk assessors to modify methods as science has evolved.[44] Prudent vigilance is required to ensure that this strategy of comparison to known relatives, when they exist, remains effective as synthetic biology advances.

Recommendation 7: Risk Assessment Prior to Field Release

Reasonable risk assessment should be carried out, under the National Environmental Policy Act or other applicable law, prior to field release of research organisms or commercial products involving synthetic biology technology. This assessment should include, as appropriate, plans for staging introduction or release from contained laboratory settings. Exceptions in limited cases could be considered, for example, in emergency circumstances or following a finding of substantial equivalence to approved products. The gap analysis described in Recommendation 5 should determine whether field release without any risk assessment is permissible and, if so, when.

This recommendation is not intended to suggest that a National Environmental Policy Act-style risks evaluation must be conducted in all cases. As noted, there are numerous models and strategies employed across the government for risk assessment, for example, through FDA's premarket and post-market processes, EPA's Toxic Substances Control Act processes, and others. The goal of this recommendation is to ensure that for any field release there is adequate consideration of risk. Through the suggested inter-agency process, the government may find that for some products—for example, first-generation fruits or vegetables developed with synthetic biology instead of traditional recombinant methods—there is no material need to establish formal risk assessment and premarket approval if not required already under existing law. Because of the uncertainty surrounding this novel technology and the great potential it presents for confusion and public fear, Recommendations 5 directs the government to affirmatively examine current policies for field release, to ensure that they are adequate, and to disclose to the public the results of this review.

The Commission's deliberations also highlighted the degree to which synthetic biology is an international enterprise. From student competitions to commercial gene synthesis companies, the synthetic biology community is an interactive global network. Oversight and regulatory mechanisms should adopt an analogous approach, so that the united States is involved in regular discussions with other national and transnational organizations, together seeking coordination and consistency when possible. These interactions should foster international collaboration as well as provide opportunities for the united States to learn from the positive and negative experiences of other countries similarly striving to promote the safe development of this field. International cooperation to create, maintain, enforce, and periodically update universal safety standards is essential.

Recommendation 8: International Coordination and Dialogue

Recognizing that international coordination is essential for safety and security, the government should act to ensure ongoing dialogue about emerging technologies such as synthetic biology. As part of the coordinated approach urged in Recommendation 4, the Executive Office of the President, through the Department of State and other relevant agencies such as the Department of Health and Human Services and the Department of Homeland Security, should continue and expand efforts to collaborate with international governments, the World Health Organization, and other appropriate parties, including international bioethics organizations, to promote ongoing dialogue about emerging technologies such as synthetic biology as the field progresses.

Creating a Culture of Responsible Stewardship

Responsible conduct of synthetic biology research, like all areas of biological research, rests heavily on the behavior of individual scientists. Federal oversight can guide the development of a culture of responsibility and accountability, but it also must be fostered at the local level. Ethical as well as biosafety and biosecurity standards are translated into practice at the laboratory level—and by the institutions that sponsor that laboratory science.[45] As an example, programs focused on homeland and transportation security embrace the message, "if you see something, say something." The same is true for laboratory science. It is at the individual or laboratory level where accidents will occur, material handling and transport issues will be noted, physical security will be enforced, and potential dual use intentions will most likely be detected.

Creating a culture of responsibility in the synthetic biology community could do more to promote responsible stewardship in synthetic biology than any other single strategy. For example, ethics education is required for most federally funded investigators conducting research with human subjects or laboratory animals.[46] Similarly, researchers working with select agents must undergo training in biosafety and biosecurity before having access to select agents and pathogens.[47] Researchers working with recombinant DNA in institutions that receive federal funds for such research know they must undergo review by an institutional biosafety committee (IBC) prior to beginning work.[48] These agreements between scientists and the public are the terms—the social contract, one might say—for conducting "risky" science, and they are well understood by most of the biological and biomedical research community.

Federal funding for engineering research, in contrast to clinical research, generally does not include a requirement for ethics training. Recently, the National Science Foundation began conditioning some research awards on agreements that institutions mentor funded postdoctoral research fellows and implement plans for "appropriate training and oversight in the responsible and ethical conduct of research. . . ."[49] Other federal research sponsors lack even these modest requirements. There is an urgent need more generally for careful consideration of the education and training necessary to promote ethical conduct in engineering research and practice.

There are new actors in the world of synthetic biology, namely engineers, chemists, materials scientists, computer modelers, and others who practice outside of conventional biological research settings.[50] These groups may not be familiar with the standards for ethics and responsible stewardship that are commonplace for those working in biomedical research. This poses a new challenge regarding the need to educate and inform synthetic biologists in all communities about their responsibilities and obligations, particularly with regard to biosafety and biosecurity.

Recommendation 9: Ethics Education

Because synthetic biology and related research cross traditional disciplinary boundaries, ethics education similar or superior to the training required today in the medical and clinical research communities should be developed and required for all researchers and student-investigators outside the medical setting, including in engineering and materials science. As part of the coordinated approach urged in Recommendation 4, the Executive Office of the President, in consultation with the National Academy of Sciences, the National Academy of Engineering, the scientific community, and the public, should convene a panel to consider appropriate and meaningful training requirements and models. This review should be completed within 18 months and the results made public.

Collectively, these recommendations are designed to balance enthusiasm for the potential benefits of synthetic biology with the vigilance required to minimize the risks associated with research in this field and its applications. Through technological and regulatory mechanisms, a spirit of international collaboration, and researcher education, the scientific and policy communities can work together to be responsible stewards for humankind, other species, and our shared environment.

Weighing Moral Objections

The Commission's discussion of synthetic biology thus far has focused on efforts to identify and assess the risks and potential benefits of research and development activities. Significant challenges exist for the scientific and regulatory communities in these areas, and these recommendations aim to strengthen systems to promote activity in this field while protecting against risks. There is a second category of concerns regarding synthetic biology, one that is largely independent of specific risk-benefit analyses related to proposed applications or research directions. These are concerns that synthetic biology is intrinsically objectionable from a moral perspective and should therefore not be allowed to proceed.[51] The term "intrinsically objectionable" is used to express the idea that an activity or practice is "bad in itself." The suggestion of some critics, moreover, is that no amount of vigilance, safeguards, or similar mechanisms could justify the transgression by synthetic biology of an important moral barrier.

Intrinsic objections have led to direct policy consequences in other areas of biomedical science and technology, most notably the restrictions on research related to human reproductive cloning and embryonic stem cell research. These types of concerns have had a long and important place in bioethical discussions and debates. Intrinsic objections to synthetic biology raise important issues deserving ongoing consideration as part of comprehensive efforts to assure that this field progresses within appropriate ethical boundaries.

The Commission learned of several possible intrinsic objections to synthetic biology during its deliberations.[52] In one formulation, synthetic biology is thought to conflict with essential concepts of human agency and life, "promoting a grandiosity about human powers or dismissiveness about the specialness of life."[53] The tools of synthetic biology and the technological capabilities they provide may, according to some critics, accentuate humankind's temptation to hubris, suggesting an expansive, even limitless, ability to shape life and the future. Related to this criticism is the suggestion that advances in synthetic biology demonstrate that life is "nothing more than the sum of its parts" and that there is nothing "unique and unknowable about life itself."[54] Contrasting synthetic biology with genetic engineering, medical ethicists Joachim Boldt and Oliver Müller write,

> [S]ynthetic biology does not soften edges, but creates life forms that are meant not to have any edges from the start. It does not add value to an existing organism; it brings into existence something that counts as valuable from our point of view. Seen from the perspective of synthetic biology, nature is a blank space to be filled with whatever we wish.[55]

Boldt and Müller argue that the transition from genetic engineering to synthetic biology marks a profound shift from the manipulation of existing species to the creation of new forms of life, a shift having considerable ethical significance. They note that the metaphors commonly used in synthetic biology which describe organisms as physical artifacts—"BioBricks," living machines, hardware and software—"may in the (very) long run lead to a weakening of society's respect for higher forms of life that are usually regarded as worthy of protection."[56]

Other commentators note that some of the potential products of synthetic biology "might fail to fit comfortably into our intuitive dichotomy between the living and the non-living."[57] For example, bacterial "bio-factories" are a potential application of synthetic biology that invokes yet another metaphor describing organisms in terms of physical artifacts. These bio-factories would possess many characteristics regularly associated with life, including a nucleic acid genome and the ability to reproduce. They would also possess features commonly associated with machines—such as modular construction and a rational design developed for specific applications.[58] Some critics of synthetic biology suggest that this amalgam of characteristics, even in single-celled organisms, could adversely affect how we understand and treat other forms of life generally, not simply those produced through synthetic biology.

Another related objection to synthetic biology is that it fails to show adequate respect for nature and the environment.[59] These critics distinguish the products of synthetic biology as unnatural in ways that other interactions between humans and nature are not.[60] Philosopher Christopher Preston writes that genomes assembled through synthetic biology "depart from a core principle of Darwinian natural selection—descent through modification."[61] He argues that synthetic biology may therefore constitute a "moral 'line in the sand.'"

Civil society organizations such as the ETC Group also express concern about the overall impact of synthetic biology on biodiversity, ecosystems, and food and energy supplies worldwide.[62] These critiques combine intrinsic moral objections to the very nature of the enterprise of synthetic biology with reservations regarding its consequences and the specific harms that may result from continued research in the field. Biodiversity, for example, could be adversely affected by unpredictable outcomes of unintentional or deliberate release of synthetic organisms. Additional harms to biodiversity could result from potential applications of synthetic biology that aim to convert "low-value" forests and agricultural products into feedstocks for energy-producing processes.[63]

Concern for the continued flourishing of plant and animal species derives from the unique ability of humans to serve as responsible stewards of nature. It also acknowledges the complex relationships that exist among species in ecosystems. Unintended consequences could result from potential synthetic biology applications that involve new or modified species in nature or novel uses for existing species.

Concerns for biodiversity are not restricted to wholesale threats to species. The potential of synthetic biology to enhance, add, or remove genes (and, therefore, proteins and their functions) within organisms highlights the potential effects of synthetic biology on genetic and genomic diversity. These impacts potentially extend also to genetic diversity among humans. Gene therapy trials using recombinant DNA in humans are already underway. However, genetic manipulation, as described above, is proceeding in limited and carefully controlled ways to potentially improve human health. The Commission is aware of no active or planned research programs involving synthetic biology applied to human genomes, which are vastly larger and more poorly understood than the bacterial genomes studied thus far.

Throughout its deliberations, the Commission took special efforts to learn the views of major faith-based communities, including those of Christianity, Judaism, and Islam. In other contexts, religious groups have expressed clear and unqualified opposition to specific scientific activities based on intrinsic arguments, such as the position of the Catholic Church on human embryonic stem cell research. Similar opposition to synthetic biology has not been voiced thus far. Following the publication of the Venter Institute's paper, an official from the Catholic Church praised the development as "a further mark of man's great intelligence, which is God's gift enabling man to better know the created world and therefore to better order it."[64] The statement encouraged continued synthetic biology research, provided that the research proceeded responsibly and did not undercut the sanctity of life.

The Commission did not hear or identify any specific objections to current research efforts in synthetic biology based on the views of organized religions. In response to claims by some commentators that the Venter Institute's research demonstrates that life is merely a manipulable series of chemical reactions without any unknowable mystery or value, the Commission heard compelling rebuttals from several faith-based thinkers and others, including many scientists. Among them, it heard that absolutely nothing accomplished in synthetic biology by way of synthesizing the genome of a self-replicating bacterial cell from its component parts—which is the most striking and specific technical achievement of the Venter Institute team—demonstrates that life is without mystery or value that goes beyond the assembly of its parts. The mystery of life is amply great, as both religious and secular minds can appreciate, to survive even the most masterful scientific feats.[65]

As a scholar from the Christian tradition commented to the Commission during its deliberations,

> The mystery of existence from a Christian theological standpoint is that anything is rather than nothing, that there is something rather than nothing. That life is possible. The dynamism and the energy of matter and being itself are taken as an expression of the very vitality of God. And neither wonder nor mystery it seems to me are vitiated by the fact that we have figured out the biomechanical and bioelectrical and biochemical mechanisms thereof.[66]

Although contemporary synthetic biology is occasionally described as "creating life," this, as a factual matter, has not happened. The field currently is capable of significant but quite limited technical achievements. Potential developments that would raise further intrinsic concerns—the synthesis of genomes for a higher order or complex species, for example—are not currently possible. There is widespread agreement that this will remain the case for the foreseeable future. Synthetic biology is currently capable of manipulation and duplication of genomes of single-celled organisms. The creation of novel, complex organisms *de novo*, the focus of some opposition to synthetic biology on intrinsic grounds, is a far more difficult technical achievement. The Commission does not find it to be an inevitable consequence of recent and ongoing research activities in synthetic biology.

After careful deliberation, the Commission was not persuaded by concerns that synthetic biology fails to respect the proper relationship between humans and nature. It was reminded during its deliberations of the challenges of defining "nature" or "natural" in this context, particularly in light of humans' long history interacting with and affecting other species, humankind, and the environment.[67]

Damaging consequences have resulted from some of this past activity. The Commission believes, however, that opposition to synthetic biology at present on such grounds alone does not adequately reflect the relationship of this technology to previous scientific activities and the current limited capabilities of the field.

These varied concerns are quite valuable, however, in calling attention to fundamental, challenging questions regarding how to best understand interactions among humans, technology, and nature beyond the limited context of synthetic biology. To what extent and in what valuable ways are the many different kinds of life on earth more than the sum of their standardized and non-standardized biological parts? Such discussions and the related attention they direct toward potential objections to synthetic biology will surely continue as the field matures, as well they should.

The question relevant to the Commission's present review of synthetic biology is whether this field brings unique concerns that are so novel or serious that special restrictions are warranted at this time.

Based on its deliberations, the Commission has concluded that special restrictions are not needed, but that prudent vigilance can and should be exercised. As this field develops and our ability to engineer higher-order genomes using synthetic biology grows, other deliberative bodies ought to revisit this conclusion. In so doing, it will be critical that future objections are widely sought, clearly defined, and carefully considered within their appropriate context.

Recommendation 10: Ongoing Evaluation of Objections

Discussions of moral objections to synthetic biology should be revisited periodically as research in the field advances in novel directions. Reassessment of concerns regarding the implications of synthetic biology for humans, other species, nature, and the environment should track the ongoing development of the field. An iterative, deliberative process, as described in Recommendation 14, allows for the careful consideration of moral objections to synthetic biology, particularly if fundamental changes occur in the capabilities of this science and its applications.

Intellectual Freedom and Responsibility

Democracies depend on intellectual freedom coupled with the responsibility of individuals and institutions to use their creative potential in morally responsible ways. Sustained and dedicated creative intellectual exploration begets much of our scientific and technological progress. A robust public policy regarding the responsible conduct of science must promote the creative spirit of scientists and unambiguously protect their intellectual freedom. At the same time, responsible science should reject the technological imperative: the mere fact that something new can be done does not mean that it ought to be done.

Society as a whole has a stake in what scientists and engineers do. In turn, scientists and engineers should recognize the potential impact of their research on those who will experience both its benefits and burdens and their responsibility to those who provide the means, directly or indirectly, for their research. As a corollary to the principle of intellectual freedom and responsibility, the Commission endorses a principle of regulatory parsimony, recommending only as much oversight as is truly necessary to ensure justice, fairness, security, and safety while pursuing the public good.

The section on responsible stewardship stressed the importance of regulatory parsimony, recommending limiting regulation to that which is necessary to promote public safety and security, public beneficence, and justice and fairness. In its discussion of democratic deliberation, the Commission recognizes the important part that all citizens can serve in working together for the common good. Responsible stewardship and democratic deliberation are two important components of a framework that promotes intellectual freedom coupled with responsibility. This section examines the central role of this principle in supporting the development of synthetic biology and other emerging technologies.

Intellectual freedom lies at the heart of America's scientific enterprise. Such freedom facilitates the innovation and industry that have fueled its success. History is rife with examples in which ingenuity, hard work, and unfettered creativity have yielded extraordinary, sometimes unexpected, scientific advances for the betterment of society as a whole. From Benjamin Franklin studying electricity with a kite in a raincloud, to the Wright Brothers testing different aerodynamic control systems and building the first successful airplane, students learn every day about the value of intellectual and scientific freedom and exploration.

Scottish scientist Alexander Fleming famously discovered the antibiotic penicillin by chance in 1928 after observing an area on a mold-contaminated Petri dish where bacteria did not grow. David Hewlett and William Packard started in their backyard garage an electronics revolution that continues to the present, working in the 1930s in what is now described as the "birthplace of Silicon Valley." And the Internet, with its vast reach today, began as a simple idea to share data among U.S. Defense Department researchers in the 1960s. These examples show that the precise outcomes of open scientific exploration and discourse cannot always be predicted, but the value they deliver as the engine of progress, in science and in society overall, is unparalleled.

Intellectual freedom and responsibility can be understood in two senses. First is the special institutional attribute—academic freedom and responsibility—that pertains to the "academy" (broadly speaking, universities and the scholars and researchers whose professional standing carries with it the rights and responsibilities of academic freedom). Some research involving synthetic biology today occurs in this setting, which includes unique institutional structures to promote the responsibility that accompanies intellectual freedom. Second is the right of all individuals to freedom of inquiry. The DIY research communities and other private researchers are exercising such freedom but without the institutional norms and procedures designed to assure responsibility, although these groups often develop their own mechanisms intended to do so.

In academic communities, intellectual freedom is essential. The ability to explore ideas openly and freely—even controversial or unpopular ideas—is fundamental to the mission of education and research. "The common good depends upon the free search for truth and its free exposition," according to one widely endorsed statement on academic freedom.[68] Academic freedom is not to be confused with license; it protects neither socially irresponsible behavior (the abuse of one's academic office) nor research that poses risks to individuals or institutions without adequate safeguards. Its limitations notwithstanding, the free exchange of ideas is essential to both academic inquiry and to the overall health of societies, and is recognized to be vital in the United States and other modern democracies. Protecting this core freedom— while meeting the corresponding responsibilities—is among the foremost concerns of academic communities.

Certain regulatory and norm-based constraints on academic and intellectual freedom in academic and other settings ensure that scientists act responsibly to protect others. In academic science, universities and other institutions accept the responsibility to abide by safety and security measures in laboratory research. These institutions, government, and most industry research programs employ extensive quality assurance and control processes that satisfy both external mandates and internal needs. In non-academic settings, like some DIY synthetic biology communities, recognition and acceptance of such processes are less common. In some cases, practitioners unaffiliated with an institution are simply unaware of applicable or reasonable restrictions governing scientific research methods intended to promote security and safety.[69]

Research Oversight Policies and Practices

Citizens and their leaders should have a voice in deciding the conditions and direction of research efforts, especially, though not exclusively, when public funds are used. Likewise, scientists have a responsibility to ensure that they use public monies wisely and act in ways consistent with public trust. Recognizing this responsibility, scientists at the early stages of the genetic engineering revolution came together to develop what remains a substantially self-regulated system to protect against physical risks in genetic research. At the historic Asilomar Conference on Recombinant DNA in 1975, scientists developed a set of principles that required containment measures to be an essential consideration in experimental design and that "the effectiveness of the containment should match, as closely as possible, the estimated risk."[70] Although the scientists recognized that it might be difficult to predict the level of risk for any particular experiment given the novel character of the research, the guidelines established graded containment strategies and categorized expected areas of inquiry, setting minimum levels of containment within the graded system. Like atomic scientists before them, the scientists who participated at Asilomar recognized that the uncertain nature of the risks associated with their efforts demanded that they act cautiously and with utmost attention to the public interest. They agreed to defer types of research that could not be carried out at that time with sufficient safeguards.

Building on this framework, scientists both in and outside government developed a shared culture of responsibility to assure safe conduct of research in the largely uncharted world of genetic engineering. In the 35 years since Asilomar, the then-nascent field of genetic engineering research has flourished. Its safety continues to be governed by a dynamic process of active engagement among scientists in academia, government, and the private sector.

Synthetic biology today finds itself in a position similar to the field of genetic engineering in 1975. Some urge extreme caution and prohibition until safety is proven, and others are perhaps too sanguine, dismissing all efforts that might limit intellectual freedom and scientific exploration. As mentioned in the sections on responsible stewardship and democratic deliberation, the Commission finds neither of these approaches appropriate. The principle of intellectual freedom and responsibility leads us to the conclusion that restrictions on research, whether by self-regulation among scientists or by government intervention, should limit the free pursuit of knowledge only when the perceived risk is too great to proceed without limit. Restrictions can prevent research harms but also can impede innovation and progress that may itself reduce harms.

In 2009, NIh recommended that synthetic biology research should be overseen at this time in the same manner as more traditional genetic engineering research. The Commission agrees. The *NIH Guidelines for Recombinant DNA Research* (the *NIH Guidelines*), discussed in Chapter 4, establish safety conditions based on the risk profile of the end product, for example, a genetically modified virus strain, rather than the techniques used to make it. Risks are assessed and safety precautions imposed based on risks, but research is not limited or restricted in the absence of realistic and identified concerns. This framework is time-tested, familiar to most researchers, and consistent with the principle of intellectual freedom and responsibility.

A moratorium at this time on synthetic biology research generally or in particular areas would inappropriately limit intellectual freedom. Instead, the scientific community—in academia, government and the private sector— should continue to work together to evaluate and respond to known and potential risks of synthetic biology as this science evolves.

Recommendation 11: Fostering Responsibility and Accountability

The government should support a continued culture of individual and corporate responsibility and self-regulation by the research community, including institutional monitoring, enhanced watchfulness, and application of the NIH Guidelines for Recombinant DNA Research. As part of the coordinated approach urged in Recommendation 4, the Executive Office of the President should evaluate, and re-evaluate periodically, the effectiveness of current research oversight mechanisms and determine what, if any, additional steps should be taken to foster accountability at the institutional level without unduly limiting intellectual freedom. Academic and private institutions, the public, the National Institutes of Health, and other federal funders of synthetic biology research should be engaged in this process. An initial assessment should be completed within 18 months and the results made public.

This activity may best be undertaken through the coordinated approach chosen to implement Recommendation 4. The Office of Science and Technology Policy or another Executive Branch office could also direct this review. The responsible office must be empowered to bring together all relevant agencies and departments and assure effective engagement with outside groups.

The notion of "enhanced watchfulness" requires the scientific community to recognize the varied risks associated with synthetic biology and develop internal processes to identify and respond to potential threats rapidly and effectively. Enhanced watchfulness reflects a relationship among scientists, citizens, and policy makers built on trust and mutual respect. To earn and preserve public trust, the research community should actively engage in continuing efforts to promote the safe development of synthetic biology and to recognize potential threats before they cause harm.

A culture of responsibility is particularly effective in university settings where academic freedom is an institutionalized right but not an unrestricted license. The responsibilities attendant to this freedom are implemented through practical mechanisms that nurture and support the culture of responsibility. Compliance with the *NIH Guidelines*, for example, is assured through a series of internal checks and balances from the investigator through to local oversight committees (e.g., IBCs) and the institutional signing official responsible for assuring that the institution meets all terms and conditions of research funding.

Researchers in institutions outside the university setting also have an incentive to limit risks and frequently have systems in place to support and sustain the culture of responsibility. Biotechnology companies staffed with scientists trained in academia and accustomed to working with oversight committees like IBCs often volunteer to comply with the *NIH Guidelines* and other standards developed through consensus of the scientific community.[71] Researchers in government agencies are also familiar with IBC review and the *NIH Guidelines*, and often are required to comply with them.

Nurturing this same culture among DIY investigators or others outside of institutional settings is more challenging. The global expansion of DIY synthetic biology raises fears about biosafety and biosecurity. The open access environment underpinning many DIY efforts, as well as the increasing affordability and availability of synthetic biology tools through private gene synthesis companies and others, generates understandable concern about the ongoing effectiveness of self-regulation and the culture of responsibility standard. In partial response, the FBI expanded efforts in the last few years to partner with industry and actively engage the DIY community on safety concerns and risk mitigation strategies.[72]

The principle of intellectual freedom and responsibility, when responsibility is exercised largely by individual rather than institutional actors, requires the government to be particularly vigilant, although perhaps no more limiting of research efforts. To exercise the appropriate level of oversight, the government will need to monitor the growth and capacity of researchers outside of institutional settings. This effort may require the government to expand current oversight or engagement activities with these non-institutional researchers. NIH or the Department of Energy, for example, could be charged to sponsor education programs and workshops that bring together these groups. They could fund training grants or related programs to promote responsibility among this community.

Recommendation 12: Periodic Assessment of Security and Safety Risks

Risks to security and safety can vary depending on the setting in which research occurs. Activities in institutional settings, may, though certainly do not always, pose lower risks than those in non-institutional settings. At this time, the risks posed by synthetic biology activities in both settings appear to be appropriately managed. As the field progresses, however, the government should continue to assess specific security and safety risks of synthetic biology research activities in both institutional and non-institutional settings including, but not limited to, the "do-it-yourself" community. As part of the coordinated approach urged in Recommendation 4, the Executive Office of the President, working with the Department of Homeland Security, the Federal Bureau of Investigation and others, should undertake and periodically update this assessment. An initial review should be completed within 18 months and the results made public to the extent permitted by law.

As above, this activity could be undertaken by a central office implementing Recommendation 4, but it need not be, provided that the implementing office has sufficient authority to accomplish this charge. The analysis recommended here should identify efforts to bring the non-institutional communities into the ongoing culture of responsibility and local accountability that currently exists in many institutional settings.

This recommendation acknowledges that the norms of safe and responsible conduct that have evolved over time for many researchers in institutional settings may not be understood or followed by those new to the field or outside of these settings, but it is not a call for specific restraints upon the DIY community at this time. Synthetic biology is occasionally critiqued as scientists "playing God," but a more general concern is ensuring that all scientists, particularly DIY scientists, reject a culture of *play* and adopt a culture of *responsibility* as it relates how they view their own research in a field fraught with risks to themselves, the public, and the environment.

It is important to note that there is presently no serious risk of completely novel organisms being constructed in non-institutional settings such as the DIY community. The research result announced by the Venter Institute in May 2010 was a significant technical achievement, but the synthesis of a self-replicating bacterial cell with a synthetic genome required nearly 15 years of work by a large team of highly experienced scientists and an estimated $40 million in research expenditures. The Commission's deliberations revealed that this combination of technical and financial resources and scientific expertise is not currently available in the DIY community. The potential synthesis of completely novel organisms presents additional, still unresolved technical challenges even for research groups working in institutional settings. While there are known risks related to near-term activities by the DIY community, such as the growth of potentially pathogenic organisms using conventional methods or inadequate waste disposal practices, the risks associated with this group using synthetic biology techniques to create novel organisms are presently quite low.

This recommendation echoes recent conclusions of the NSABB, which also considered issues of education and outreach to all practitioners of synthetic biology and ways to effectively promote a culture of responsibility.[73] Scrutiny is required to assure that DIY scientists have an adequate understanding of necessary constraints to protect public safety and security, but at present the Commission sees no need to impose unique limits on this group.

Assessing Oversight and Export Controls

The culture of responsibility depends, at least in part, on voluntary compliance with the *NIH Guidelines* in institutions without federal research funding, such as private companies. Accordingly, the Commission recommends that the government undertake an ongoing process of review to monitor risks and effectiveness of current oversight systems in these settings and in contexts such as the DIY community.

However, certain risks—generally involving national security—often warrant additional protections. One of the primary concerns about the risks posed by synthetic biology is its dual use potential, defined as the possibility that it will yield information or technologies capable of being misused, thereby endangering public health or national security.

The threat of malevolent use of scientific knowledge is not new; however, the global, collaborative, and electronically linked nature of modern biological sciences, such as synthetic biology, complicates efforts to control scientific information and material exchanges across borders.

Where uncertainty exists regarding the danger of specific genetic sequences that potentially code for harmful substances, sequence providers should strive to ensure that customers and end-users have legitimate purposes for their use. Adherence to the government's voluntary *Screening Framework Guidance for Providers of Synthetic Double-Stranded DNA* will aid these efforts.[74]

Scientists and laboratory technicians should ensure that containment and other safety precautions are in place. The scientific community should take steps to carefully manage both scientific and social risks associated with synthetic biology as this field grows.

Chapter 4 briefly describes the current system of export controls and other measures designed to reduce concerns about malevolent use arising from information exchange. Policy makers in this area face complex challenges. Completely free exchange of data and materials might endanger public safety, but unilateral action to limit exchange could damage American research efforts in synthetic biology if U.S. scientists and students are excluded from full collaboration in the international community.

An additional complication for export control efforts in synthetic biology is that much of the "currency" of the field are the sequences of genetic data that are often available in public databases or could be distributed easily and without detection.

Several recent advisory groups have recommended ongoing discussions among research universities, industry, and government on this topic. The National Research Council's 2007 report, *Science and Security in a Post 9/11 World*, expressly calls for more dialogue on export controls. The NSABB in 2010 also recommended expanded outreach and education strategies "that address dual use research and engage the research communities that are most likely to undertake work under the umbrella of synthetic biology."[75] The Commission agrees that scientists should be actively engaged in these debates.

Recommendation 13: Oversight Controls

If the reviews called for in Recommendation 12 identify significant unmanaged security or safety concerns, the government should consider making compliance with certain oversight or reporting measures mandatory for all researchers, including those in both institutional and non-institutional settings, regardless of funding sources. It may also consider revising the Department of Commerce's export controls. Any such change should be undertaken only after consultation with the scientific, academic, and research communities and relevant science and regulatory agencies such as the National Institutes of Health, the Department of Homeland Security, and the Environmental Protection Agency. Export controls should not unduly restrain the free exchange of information and materials among members of the international scientific community.

Democratic Deliberation

The principle of democratic deliberation reflects an approach to collaborative decision making that embraces respectful debate of opposing views and active participation by citizens. At the core of democratic deliberation is an ongoing, public exchange of ideas, particularly regarding the many topics—in science and elsewhere—in which competing views are advocated, often passionately. A process of active deliberation and justification promotes an atmosphere for debate and decision making that looks for common ground wherever possible, and seeks to cultivate mutual respect where irreconcilable differences remain. It encourages participants to adopt a societal perspective over individual interests. With careful attention to the processes through which decisions are reached and justified, democratic deliberation promotes outcomes that are inclusive, thoughtfully considered, and respectful of competing views.

Biotechnology has the potential to affect everyone, and opportunities for the public to participate in discussion and deliberation about emerging technologies such as synthetic biology are critical. The principle of democratic deliberation highlights the importance of robust public participation in both the development and implementation of specific policies as well as in a broader, ongoing national conversation about science, technology, society, and values.

In its examination of synthetic biology, the Commission saw encouraging examples of ways in which the public has been invited to learn about this emerging field and to share its perspectives. It learned of groups of citizens coming together, sharing their mutual interest and expertise in synthetic biology—biologists and engineers, teachers and students, professionals and amateurs, from nations around the world. These activities provide an important foundation for expanded efforts regarding public engagement and public education that are not only valuable but essential. This section highlights examples of how citizens are already shaping the present and future of synthetic biology and notes several opportunities for how these efforts can be enhanced and strengthened.

Promoting an Ongoing Public Dialogue

Many groups in addition to this Commission have studied and reported on issues related to synthetic biology in the past several years, including u.S. and international government agencies, professional societies, commercial and industry groups, and private organizations. As the Commission did throughout its deliberations, virtually all of these groups consulted broadly among those with interest and expertise regarding the potential impact of synthetic biology on science and society. The Commission commends these efforts, as they embody a belief that policy regarding synthetic biology is best developed when informed by open and ongoing discussions among a diverse group of stakeholders. Policymaking bodies involved in regulation and oversight of synthetic biology are encouraged to continue to actively solicit input from the public regarding their work, ensure that those views receive thoughtful consideration, and make available and accessible to the public the eventual decisions that are reached and the reasoning for them. Public deliberation is particularly valuable while the field is still young, as there is a unique opportunity to shape its development in ways most likely to promote the public good while assuring safety and security.

The Commission understands that not all policymaking activities in this area can be fully transparent to the public, such as those related to some aspects of biosecurity or involving trade secrets in certain commercial applications of synthetic biology. Concerns about biosecurity and proprietary interests ought not, however, justify excessive secrecy such that the development of science and the participation of the public are unduly compromised. Nor should these necessary limitations preclude those with advisory or decision-making responsibilities from viewing the public as active partners in their work. In addition to being a valuable source of good ideas, public participation frequently fosters the perceived political legitimacy of the policies and practices that are ultimately chosen.

A recent survey of public attitudes regarding synthetic biology found that nearly two-thirds of respondents supported continued development of the field, including additional research on its possible effects on humans and the environment.[76] There was a strong correlation between self-reported awareness of synthetic biology and support for ongoing research, as 80 percent of those who had heard a lot about the field believed it should move forward, compared to only 52 percent of those who had heard nothing about it. Overall, 73 percent of those surveyed reported having heard "just a little" or "nothing at all" about synthetic biology. These data indicate both the need for broader public engagement regarding synthetic biology and the positive impact of such efforts on public support for novel and otherwise unfamiliar technologies.

In many areas of biomedical research, public engagement is an important component of study design and a means to ensure public support. A notable example of this practice is the Framingham Heart Study and its Ethics Advisory Board. The study, which began in 1948, is a federally funded project based in Framingham, Massachusetts that aims to identify and understand the risk factors for heart disease by observing entire families and populations over time. The Framingham Ethics Advisory Board is comprised largely of past and present study participants as well as local clergy and physicians. It serves as a forum for community deliberations and a vehicle to advise the researchers on design and oversight issues.[77]

The development of the NIH policy on Genome-Wide Association Studies demonstrates another type of proactive public engagement to build public understanding and support. In connection with building a large-scale, central database of individual genotype and phenotype information for secondary research studies, NIH published requests for public comment during the policy development process and held meetings with members of the public prior to finalizing its policy.[78] In another example, community engagement is required by law for certain research projects in which individual informed consent is not feasible, such as research conducted in emergency settings.[79] Increasingly, community engagement or consultation is a prerequisite for research with particular populations, such as Native Americans, or research requiring the use of high-containment facilities to control dangerous pathogens.

Other groups have noted the potential value of public engagement specifically for synthetic biology and related topics. In its April 2010 report on synthetic biology, NSABB recommended outreach and education directed toward participating scientific communities, while also stating that more active engagement of the general public could lead to a better collective understanding of synthetic biology.[80]

An active, dynamic exchange between citizens and government need not be confined to regulatory and legislative processes. During its deliberations the Commission learned of several initiatives in which government agencies such as the FBI are in regular dialogue with members of the synthetic biology community.[81] These activities provide opportunities for citizens and their government to learn from each other, exchange ideas, share concerns, and work collaboratively toward fostering a safe, productive environment in which synthetic biology can develop.

Government bioethics commissions such as this one can be part of national and international conversations regarding synthetic biology and other emerging technologies.[82] While by no means a substitute for robust, ongoing exchanges between citizens and policy makers, the Commission's deliberations on this matter sought to provide an inclusive forum for discussion, with the hope that its recommendations will be a catalyst for future deliberations.

The Commission's interest in democratic deliberation calls for a national and international dialogue on synthetic biology and its implications, a conversation that bridges specific research initiatives and considers how the field as a whole can best move forward safely and beneficially. The Public Engagement with Research Team of the Research Councils uK is one example of an approach that promotes sustained interactions among researchers, students, and the public on major themes related to research and innovation.[83]

Recommendation 14: Scientific, Religious, and Civic Engagement

Scientists, policy makers, and religious, secular, and civil society groups are encouraged to maintain an ongoing exchange regarding their views on synthetic biology and related emerging technologies, sharing their perspectives with the public and with policy makers. Scientists and policy makers in turn should respectfully take into account all perspectives relevant to synthetic biology.

Democratic deliberation encourages respect for a wide range of reasonable perspectives. Positions based directly on personal revelations—whether divine or secular in nature—are unlikely to be accessible to most citizens. However, by carefully attending to the concerns raised by religious traditions, a respectful dialogue can develop that can often lead to positions that are accessible, independent of their source.[84] While the Commission did not observe significant religious concerns related to synthetic biology at this time, the field is young, and future developments may prompt new concerns, underscoring the importance of ongoing deliberation that is responsive to changing circumstances in science and society.

Striving for Accuracy and Understanding

For effective public deliberation on potentially contentious topics such as synthetic biology, participants should endeavor to express their views in ways that are accessible to others. In part, this means striving to convey one's own views and those of others accurately and with as much mutual understanding as possible. Throughout its deliberations, the Commission was impressed by the quality of discourse on synthetic biology from those working in and around the field. It did observe, however, that the media sometimes described synthetic biology in ways more provocative than accurate. This observation may not be surprising to some, but it makes the development of ongoing deliberative forums on science all the more essential to enhancing public understanding.

In the days immediately following the May 20, 2010, announcement of the creation of the first self-replicating cell containing an entirely synthetic genome, some press accounts worldwide declared, "Scientists have created the world's first synthetic life form."[85]

Subsequent coverage attempted to place this work in context, particularly regarding whether it could properly be described as truly creating synthetic or artificial life. In its deliberations, the Commission heard that while the May 20 announcement marked a significant technical achievement in demonstrating that a relatively large genome could be accurately synthesized and substituted for another, it did not amount to the "creation of life"[86] (see Chapters 2 and 3 for further discussion).

While this interpretation of the research appears to be widely held among the scientific community, public perceptions of synthetic biology may have been influenced by initial news of "creating life." This language may excite public interest in a potentially transformative field, but it can serve a useful purpose only if it is followed by careful and robust deliberation informed by an accurate understanding of the current state of synthetic biology and the uncertainty regarding its potential benefits and risks. This example illustrates the considerable opportunities and challenges facing science journalists today to excite public interest and convey accurate understanding of developments in science and technology.

Discussions about synthetic biology and related technologies often raise objections that scientists are "playing God." The Commission's deliberations with representatives of a range of religious communities found this language to be unhelpful at best, misleading at worst. It learned that secular critics of the field are more likely to use the phrase "playing God" than are religious groups. While religious thinkers suggested caution regarding the human tendency toward hubris, none expressed concern that synthetic biologists were "playing God."[87] The provocative nature of this phrase does more to obscure rather than to illuminate those important moral concerns regarding synthetic biology that deserve serious consideration.

Recommendation 15: Information Accuracy

When discussing synthetic biology, individuals and deliberative forums should strive to employ clear and accurate language. The use of sensationalist buzzwords and phrases such as "creating life" or "playing God" may initially increase attention to the underlying science and its implications for society, but ultimately such words impede ongoing understanding of both the scientific and ethical issues at the core of public debates on these topics. To further promote public education and discourse, a mechanism should be created, ideally overseen by a private organization, to fact-check the variety of claims relevant to advances in synthetic biology.

Public deliberation about synthetic biology can be hindered both by imprecise language such as "creating life" or "playing God" as well as by scientific claims that fail to convey accurately to the public the current state of the field, the implications of research results, and the limits of scientists' present knowledge and abilities. The fact-check mechanism recommended here is intended to address these concerns by providing an independent venue where scientific claims related to synthetic biology or other emerging technologies are evaluated by impartial, qualified experts. The results of these analyses would be readily accessible to the public, likely through a website. The Commission envisions a program analogous to FactCheck.org, a project that monitors the accuracy of statements made about U.S. politics.[88] It would be interactive, inviting the public to suggest claims for review by project staff, and funded by private sources without real or perceived conflicts of interest.

Improving Scientific and Ethical Literacy

Meaningful citizen participation in deliberations regarding synthetic biology requires familiarity with general concepts in science and particular aspects of this developing field. Collectively, these tools are referred to as "scientific literacy."[89] The National Academy of Sciences has defined scientific literacy as "the knowledge and understanding of scientific concepts and processes required for personal decision making, participation in civic and cultural affairs, and economic productivity."[90]

Making science accessible to the public requires creativity and innovation in public education. The Commission was pleased to learn that in synthetic biology several groups have launched commendable efforts to educate the public about this emerging field. These groups include the Synthetic Biology Project of the Woodrow Wilson International Center for Scholars and the Synthetic Biology Engineering Research Center, which is funded in part by the National Science Foundation.[91] Through online resources, curricula for teachers and students, and events such as the International Genetically Engineered Machine (iGEM) competition (see p. 46), these and other groups are developing innovative programs to increase the public's understanding of synthetic biology.

Public education efforts addressing synthetic biology need to be part of our Nation's expanded attention to an increasingly urgent need to enhance scientific literacy, broadly understood. Scientific literacy must go hand-in-hand with improved ethical literacy, meaning an understanding of moral concepts, traditions, and controversies concerning the responsibilities and rights of individuals and communities toward one another.

Recommendation 16: Public Education

Educational activities related to synthetic biology should be expanded and directed to diverse populations of students at all levels, civil society organizations, communities, and other groups. These activities are most effective when encouraged and supported by various sources, not only government, but also private foundations and grassroots scientific and civic organizations. As part of the coordinated approach urged in Recommendation 4, the Executive Office of the President, with input from the scientific community, the public, and relevant private organizations, should identify and widely disseminate strategies to promote overall scientific and ethical literacy, particularly as related to synthetic biology, among all age groups.

This effort could be led by EOP or the relevant science agencies such as NIh or DOE in collaboration with the Department of Education. This group should consider the feasibility of including public education components or the development of school curriculum modules in research funding agreements. These activities could be linked to specific projects or organized at the institutional level among recipients of federal research support. It should also examine other models to promote and enhance scientific and ethical literacy, including activities directed by private organizations or developed by private groups in partnership with the government. The Synthetic Biology Project of the Woodrow Wilson Center may serve as one such model.

Scientific research and public education about science are best approached as mutually related, even mutually dependent, endeavors. The iGEm competition, for example, combines hands-on student exposure to research tools and practices with education on many issues—science, safety, and policy—related to synthetic biology. The Commission commends programs throughout the scientific community that include educational programs as components of ongoing research projects. One illustrative example is Project BioEYES, which provides classroom-based learning opportunities for students in grades K-12 through the use of live zebrafish.[92] With active participation from scientists committed to making science accessible to young people, over 18,000 students in Philadelphia, Baltimore, and South Bend have encountered science in innovative ways. In particular, this project and others similarly directed to under-resourced schools seek to make science available to all students, particularly those who might otherwise lack access to cutting-edge scientific resources and expertise.

In 1999, the National Bioethics Advisory Commission noted that the need for expanded education is "not simply...the provision of information with the aim of adding to the net store of knowledge by any one person or group; rather, education refers to the ongoing effort to inform, challenge, and engage."[93] Engaging citizens—and particularly young people—in challenging science curricula regarding synthetic biology and other emerging technologies as well as many other issues lies at the intersection of science and citizenship. In light of our Nation's dependence on socially responsible scientific innovation for economic progress and individual well-being, the urgency of expanding effective science and ethics education cannot be exaggerated.

Fostering Grassroots Collaborations

As noted, democratic deliberation is based on ongoing interaction among citizens on topics of common interest. For an emerging technology such as synthetic biology, many of these dialogues will be among scientists or other interested citizens and policy makers or regulators. Such interactions are vital to a democracy, but they are not sufficient. Exchanges among individuals and groups of citizens are also important. In particular, grassroots collaborations have been established around synthetic biology. Groups such as DIYbio are loosely organized networks of self-described "citizen scientists" coming together because of a common interest in the tools, methods, and applications of synthetic biology, rather than shared professional affiliations or policy responsibilities. In this way, the "do-it-yourself" community embodies a "do-it-together" ethos.[94]

These kinds of collaborations are commendable; they strengthen notions of citizenship and community at the core of a democracy. They demonstrate that science and its oversight do not belong exclusively to experts, highly trained professionals, or government officials. Science is a shared resource, affecting and belonging to all citizens.

Through democratic deliberation, questions raised by the emerging science of synthetic biology can be explored and evaluated on an ongoing basis in a manner that welcomes the respectful exchange of opposing views. This deliberation is best positioned to succeed when it includes a diverse set of accessible arguments built upon a foundation of public understanding and engagement with science and technology. In this way, democratic deliberation advocates for an inclusive view of synthetic biology and its oversight. A community-oriented perspective strengthens efforts to ensure that this science develops in ways that will be acceptable to the majority of the population. This perspective also complements activities intended to promote justice and fairness in the development of synthetic biology and its applications.

Justice and Fairness

The principle of justice and fairness relates to the distribution of benefits and burdens across society. Emerging technologies like synthetic biology, for good or ill, affect all persons. Society as a whole has a claim toward reasonable efforts on the part of both individuals and institutions to avoid unjust distributions of the benefits, burdens, and risks that such technologies bring. This same claim extends internationally to all those who may be affected—positively or negatively—by synthetic biology and its applications.

In calling attention to justice and fairness, the Commission highlights the importance of considering not simply what the benefits and risks of synthetic biology are, but to whom and to what those benefits and risks are directed. Its examination of synthetic biology discussed strategies intended to realize potential benefits and minimize risks by means of thorough, inclusive deliberative processes. These benefits and risks can be specified; they are not abstract concepts. They have the potential to directly and significantly affect individuals and entire populations, species, and environments. The principle of justice and fairness encourages a proactive sensitivity to the distribution of these outcomes.

Justice and fairness are concepts with closely related but distinct meetings. Many definitions exist, but justice is generally the broader concept of the two. One type of justice, distributive justice, refers to concern for the equitable allocation of goods and evils in a society. Fairness provides one specification of justice, as in philosopher John Rawls's principles of equal liberty and equal opportunity among members of communities as two of the primary attributes of "justice as fairness."[95] In its work, the Commission refers to the principles of justice and fairness collectively to refer broadly to concern for how benefits and burdens ought to be shared among communities and nations.

Some of the most exciting potential current applications of synthetic biology involve products with the potential to address major challenges in global health and welfare. Semi-synthetic artemisinin (see p. 65), for example, could offer a valuable treatment for malaria around the globe. Synthetic biofuels could be particularly valuable in nations where energy deficits hinder development and economic growth. There is great value in striving to pursue these and other applications and to ensure, if successful, that they reach those individuals and communities who would most benefit from them.

As it advances, synthetic biology may also pose a spectrum of risks to human health, other species, ecosystems, and national security (see Chapter 3). The likelihood and severity of most of these risks are difficult to predict at this time, but part of the work of oversight activities, broadly speaking, is to assess where the risks and harms of synthetic biology are most likely to be experienced, if at all, and act to prevent or minimize any adverse impacts. Research-related risks, potential environmental exposures, and social and economic displacement can be unavoidable hazards of science and technology, but these burdens should not fall disproportionately on any particular individual or group. Of great concern are those individuals and groups whose political, economic, or other status makes them particularly vulnerable.

Sensitivity to the fair distribution of the risks and benefits of synthetic biology, like other biotechnologies, is appropriate regardless of the source of funding. yet fair distribution of the benefits of synthetic biology is an especially important consideration for government-funded research. Government support provides both benefits and obligations. Benefits include the creation of a safe and secure research environment as well as direct funding for particular projects. These benefits come with a corresponding responsibility for beneficiaries to do their part to ensure that return on these investments is justly distributed across society. Concern for justice and fairness should be a central consideration of all aspects of the planning and implementation of research in synthetic biology and its applications.

Just Distribution of Risks, Burdens, and Potential Benefits

With any technological advance come burdens and risks. These burdens can arise both in the research and development process and from the eventual introduction of new technologies and products into the marketplace. Frequently, the risks are unknown or of uncertain magnitude at the early stages of development of a field. Ongoing and recurring risk assessment is often required to fully understand and respond appropriately.

Chapter 3 discusses the potential benefits and risks of synthetic biology as it is understood today. One set of risks relates to the conduct of synthetic biology research. These include risks to laboratory workers and personnel, risks to research subjects, and risks related to the unintentional or deliberate release of experimental agents into the environment. In the United States, numerous oversight systems are in place to guard against potential harms that may result from these types of risks (see Chapter 4). These include provisions designed to prevent physical harm to workers, study subjects, and the public generally.

For study subjects, specific mechanisms are in place to ensure that volunteers are fully informed about, and agree to accept, the possible risks or harms they may face before they begin. Oversight bodies assess research risks in light of potential benefits to individuals, and in some cases, communities, prior to approval. Many believe also that research should be responsive to the needs of the entire population being studied or affected by the research activities.[96]

Evaluation of research proposals and ongoing review should include consideration of possible environmental exposures or social disruption. These considerations are particularly relevant for synthetic biology. Clinical and observational research in this field is relatively limited at present, but harmful environmental effects or unintended consequences on human health loom as major sources of concern and public anxiety. These concerns need not be addressed by institutional review boards, which are commonly understood to be prohibited by federal regulation from considering such effects beyond their relevance to the protection of human subjects directly participating in the research.[97] To address the uncertain or potentially unique risks that may arise from synthetic biology in light of its extraordinary potential to manipulate and manage living systems, special consideration and safety reviews may be needed. In addition to concerns about possible environmental exposures or social disruption, consideration must also be given to potential hazards to the public posed by synthetic biology consumer products, including medicines.

Recommendation 17: Risks in Research

Risks in research should not be unfairly or unnecessarily borne by certain individuals, subgroups, or populations. As part of the coordinated approach urged in Recommendation 4, the Executive Office of the President should lead an interagency evaluation of current requirements and alternative models to identify mechanisms that ensure that the risks of research in synthetic biology, including for human subjects and other affected parties, are not unfairly or unnecessarily distributed. Relevant scientific, academic, and research communities, including those in the private sector, should be consulted. This review should be completed within 18 months and the results made public.

Attention to these concerns is particularly relevant when those participating directly in research or likely to be affected by research activities do not share the nationality, culture, economic status, or political power of those conducting the research.

The introduction of new technologies may also lead to increased risk of harmful environmental exposures in specific locations, and the principles of justice and fairness require vigilant attention to these environmental risks. The arrival of new products or applications of synthetic biology should not compel any particular population to "shoulder a disproportionate burden of the negative human health and environmental impacts of pollution or other environmental hazards."[98] All citizens ought to "enjoy the same degree of protection from environmental and health hazards."[99] Accordingly, the Commission makes the following recommendation as a means to expand attention to the relative burden that some communities or individuals may bear regarding the potentially adverse effects and risks of new technologies.

Recommendation 18: Risks and Benefits in Commercial Production and Distribution

Risks to communities and the environment should not be unfairly distributed. Manufacturers and others seeking to use synthetic biology for commercial activities should ensure that risks and potential benefits to communities and the environment are assessed and managed so that the most serious risks, including long-term impacts, are not unfairly or unnecessarily borne by certain individuals, subgroups, or populations. These efforts should also aim to ensure that the important advances that may result from this research reach those individuals and populations who could most benefit from them. As part of the coordinated approach urged in Recommendation 4, the Executive Office of the President should evaluate current statutory mandates or regulatory requirements for distribution of risks and benefits and consider developing guidance materials and voluntary recommendations to assist manufacturers as appropriate.

There is considerable enthusiasm among advocates of synthetic biology for the varied benefits that this emerging field may yield for individuals and communities. Some critics have expressed concern, however, that synthetic biology will only exacerbate existing disparities with regard to health, welfare, and socioeconomic status.[100] Similar concerns are often voiced in response to other new technologies.

Much of the optimism surrounding synthetic biology stems directly from its potential to address some of the longstanding, significant problems associated with these disparities. Synthetic biology offers potential applications that may be particularly beneficial to less advantaged populations, including improved quality and access to vaccines against infectious diseases, medications, and fuel sources. A just society recognizes the value of establishing incentives to create new knowledge and to translate it into vibrant markets in ways intended to distribute benefits widely. As new tools arrive and mature, it will be important to identify strategies to responsibly ensure that communities and nations who may most immediately benefit are empowered to do so. Doing so will require ongoing review of how intellectual property and licensing arrangements can best be structured to promote both scientific innovation and the public good.

Stakeholders should work collaboratively, aiming to ensure that advances made possible by synthetic biology reach those who could benefit from them, particularly less advantaged populations. Attention to the just distribution of potential benefits is most effective when continually examined in concert with research and development activities. It encourages an awareness of the full "life cycle" of a new application of synthetic biology, from initial research through potential global implementation. This holistic perspective recognizes that decisions made even in early stages of development may have consequences—technological, economic, or practical—that can affect the eventual implementation of potential research products positively or negatively. The ongoing development of semi-synthetic artemisinin is an example of a research program that reflects an appreciation for the challenges and importance of ensuring wide access to possible products.[101] Research and development activities throughout synthetic biology would be well served by similar appreciation of the relationships among current activities, potential future implementation concerns, and the concepts of justice and fairness.

APPENDIX

Guest Speakers

Bonnie L. Bassler, Ph.D.
Howard Hughes Medical Institute
Investigator; Squibb Professor,
Department of Molecular Biology,
Princeton University;
President,
American Society for Microbiology

Jason Bobe, M.S.I.S.
Director of Community, Personal
Genome Project, Harvard Medical
School; Co-Founder, DIYbio.org

Sydney Brenner, M.D., D.Phil.
Senior Distinguished
Fellow of the Crick-Jacobs Center,
The Salk Institute

Allen Buchanan, Ph.D.
James B. Duke Professor of Philosophy;
Investigator, Institute for Genome
Sciences and Policy, Duke University;
Distinguished Research Associate,
Uehiro Centre for Practical Ethics,
University of Oxford

Arthur L. Caplan, Ph.D.
Emmanuel and Robert Hart
Director, Center for Bioethics;
Sydney D. Caplan Professor
of Bioethics, School of Medicine,
University of Pennsylvania

Alexander M. Capron, LL.B.
University Professor; Scott H. Bice
Chair in Healthcare Law, Policy
and Ethics; Professor of Law,
Keck School of Medicine;
Co-Director, Pacific Center for
Health Policy and Ethics,
University of Southern California

Robert Carlson, Ph.D.
Principal, Biodesic

George Church, Ph.D.
Professor of Genetics, Harvard Medical School
James J. Collins, Ph.D.
University Professor;
William F. Warren Distinguished
Professor; Professor of Biomedical
Engineering; Co-Director,
Center for BioDynamics,
Boston University; Investigator,
Howard Hughes Medical Institute

Drew Endy, Ph.D.
Terman Fellow & Assistant Professor
of Bioengineering, Stanford University;
Director, BIOFAB: International Open
Facility Advancing Biotechnology;
President, The BioBricks Foundation

Ruth R. Faden, Ph.D.
Director, Johns Hopkins Berman
Institute of Bioethics; Philip Franklin
Wagley Professor of Biomedical Ethics;
Professor, Department of Health Policy
and Management, Bloomberg School
of Public Health; Professor, Department of
Medicine, Johns Hopkins University

Gregory Kaebnick, Ph.D.
Research Scholar, The Hastings Center;
Editor, The Hastings Center Report

Nancy M.P. King, J.D.
Professor, Department of Social Sciences
and Health Policy, Wake Forest University
School of Medicine; Co-Director,
Wake Forest University Center for
Bioethics, Health, and Society

Ingrid Mattson, Ph.D.
Director, Macdonald Center for
the Study of Islam and Christian-Muslim Relations;
Director, Islamic Chaplaincy Program;
Professor of Islamic Studies and
Christian-Muslim Relations,
Hartford Seminary; President, Islamic Society
of North America

Jonathan D. Moreno, Ph.D.
David and Lyn Silfen University
Professor; Professor, History and
Sociology of Science; Professor,
Philosophy; Professor of Medical Ethics,
School of Medicine,
University of Pennsylvania

Thomas H. Murray, Ph.D.
President, The Hastings Center

Bryan G. Norton, Ph.D.
Distinguished Professor of Philosophy,
School of Public Policy,
Georgia Institute of Technology

Amy Patterson, M.D.
Acting Director, Office of Science
Policy; Director, Office of Biotechnology
Activities, National Institutes of Health

Kristala L.J. Prather, Ph.D.
Assistant Professor, Department
of Chemical Engineering,
Massachusetts Institute of Technology

Arti K. Rai, J.D.
Elvin R. Latty Professor of Law,
Duke University School of Law;
Member, Institute for Genome Science
and Policy,
Duke University

David Rejeski
Director, Science and Technology
Innovation Program, Woodrow Wilson
International Center for Scholars

David A. Relman, M.D.
Thomas M. and Joan C. Merigan
Professor of Medicine; Chief,
Division of Infectious Diseases,
Department of Medicine; Professor,
Department of Microbiology and Immunology,
Stanford University School of Medicine;
Chief of Infectious Diseases,
Veterans Administration Palo Alto
Health Care System; Chair,
Working Group on Synthetic Biology,
National Science Advisory
Board for Biosecurity

Randy D. Rettberg, M.S.
Director, iGEM and MIT Registry
of Standard Biological Parts;
Principal Research Engineer,
Department of Biological Engineering,
and Computer Science and
Artificial Intelligence Laboratory,
Massachusetts Institute of Technology

Michael Rodemeyer, J.D.
Lecturer, Department of Science,
Technology & Society,
School of Engineering and Applied Science,
University of Virginia

Markus Schmidt, Ph.D.
Co-founder, Organisation for International
Dialogue and Conflict Management,
Vienna, Austria

Allison Snow, Ph.D.
Professor, Department of Evolution,
Ecology and Organismal Biology,
Ohio State University

Ashley J. Stevens, D.Phil.
Special Assistant to the
Vice President of Research,
Office of Technology Development;
Senior Research Associate, Institute for
Technology Entrepreneurship and Commercialization,
Boston University; President,
Association of University Technology Managers

Damon A. Terrill, J.D., M.A.
Senior Vice President and General
Counsel for International Legal and
Regulatory Affairs, Integrated DNA Technologies;
Member, International Gene Synthesis Consortium

Jim Thomas
Programme Manager, ETC Group

J. Craig Venter, Ph.D.
Founder and President,
J. Craig Venter Institute

Ralf Wagner, Ph.D.
Chief Executive Officer and
Chief Science Officer,
Geneart AG; Professor and Chief,
Molecular Microbiology and Gene Therapy,
Institute of Medical Microbiology &
Hygiene University of Regensburg;
Member, International Gene
Synthesis Consortium

David B. Weiner, Ph.D.
Professor, Department of Pathology and
Laboratory Medicine; Chair,
Therapy and Vaccines Program, CAMB,
University of Pennsylvania; Chair,
Scientific Advisory Board,
Inovio Pharmaceuticals

Ron Weiss, Ph.D.
Associate Professor,
Department of Biological Engineering
and Department of Electrical
Engineering and Computer Science,
Massachusetts Institute of Technology

Sondra E. Wheeler, Ph.D.
Martha Ashby Carr Professor of
Christian Ethics,
Wesley Theological Seminary

Hugh Whittall
Director, Nuffield Council on Bioethics

Paul Root Wolpe, Ph.D.
Asa Griggs Candler Professor of Bioethics;
Director, Center for Ethics,
Emory University

Edward H. You
Supervisory Special Agent,
Federal Bureau of Investigation,
Weapons of Mass Destruction Directorate,
Countermeasures Unit I, Bioterrorism Prevention Program

End Notes for Introduction

[1] Gibson, D.G., et al. (2010). Creation of a bacterial cell controlled by a chemically synthesized genome. *Science* 329(5987):52-56.

[2] *See, e.g.*, Endy, D., Terman Fellow and Assistant Professor of Bioengineering, Stanford University. (2010). Overview and Context of the Science and Technology of Synthetic Biology. Presentation to the Presidential Commission for the Study of Bioethical Issues, July 8, 2010. Available at: http://bioethics.gov/transcripts/ synthetic-biology/070810/overview-andcontext-of-the-science-and-technology. html; Prather, K., Assistant Professor, Department of Chemical Engineering, Massachusetts Institute of Technology. (2010). Applications of Synthetic Biology. Presentation to the Presidential Commission for the Study of Bioethical Issues, July 8, 2010. Available at: http://bioethics.gov/transcripts/ synthetic-biology/070810/ applications-of-synthetic-biology.html; Petsko, G.A. (2010). Hand-made biology. *Genome Biology* 11:124.

[3] Letter from President Barack Obama to Dr. Amy Gutmann, Chair, Presidential Commission for the Study of Bioethical Issues, May 20, 2010. Available at: http://bioethics.gov/ documents/Letter-from-President-Obama-05.20.10.pdf.

[4] National Academies, Board on Life Sciences. (2010). *Sequence-Based Classification of Select Agents: A Brighter Line*. Washington, D.C.: National Academies Press; National Science Advisory Board for Biosecurity. (2010). *Addressing Biosecurity Concerns Related to Synthetic Biology*. Available at: http://oba.od.nih.gov/biosecurity/ biosecurity_ documents. html.

[5] *The Belmont Report*, issued in 1979 by the National Commission for the Protection of Human Subjects of Biomedical and Behavioral Research, was pathbreaking in the field of bioethics. It articulated and applied a three-pronged ethical framework: respect for persons, beneficence, and justice. These principles have become a widely used tool through which to assess questions in human subjects research and related topics. However,

the *Belmont Report's* principles demand refinement in light of new knowledge, different circumstances, and changing experiences. By proactively reflecting on and applying the ethical principles most relevant to our actions, we pursue a practical wisdom—knowledge that aims to produce the best results for society and the world.

[6] The Commission contrasts its work in this respect with that of the National Bioethics Advisory Commission, which was asked by President William Clinton to review ethical issues related to reproductive cloning *after* the successful cloning of an adult sheep. *See* National Bioethics Advisory Commission (1997). *Cloning Human Beings*. Available at: http://bioethics.georgetown.edu/pcbe/reports/past_commissions/nbac_cloning.pdf.

[7] National Commission for the Protection of Human Subjects of Biomedical and Behavioral Research. (1979). *The Belmont Report: Ethical Principles and Guidelines for the Protection of Human Subjects of Research*. Available at: http://www.hhs.gov/ohrp/humansubjects/ guidance/belmont.htm.

[8] For example, the "Variola Amendment" to the Intelligence Reform and Terrorism Prevention Act of 2004 made it criminal to produce, engineer, or synthesize the variola virus. The "variola virus" was in turn defined to include "any derivative of the variola major virus that contains more than 85 percent of the gene sequence of the variola major virus or the variola minor virus." 18 U.S.C. § 175c. The broad definition of "variola virus" made it unclear what was actually covered by the statute, and a strict interpretation of the definition would have potentially, and inadvertently, criminalized beneficial research such as production of the smallpox vaccine.

[9] Gutmann, A., and D. Thompson. (1996). *Democracy and Disagreement*. Cambridge, Mass.: Belknap Press; Gutmann, A., and D. Thompson. (1997). Deliberating about bioethics. *The Hastings Center Report* (May/June):38-41; Gutmann, A., and D. Thompson. (2000). Why deliberative democracy is different. *Social Philosophy and Policy* 17:161-180; Gutmann, A., and D. Thompson. (2002). Just deliberation about health care. In *Ethical Dimensions of Health Policy*. M. Danis, C. Clancy, and L. Churchill (eds.). Oxford: Oxford University Press.

[10] Daniels, N. (2008). *Just Health: Meeting Health Needs Fairly*. Cambridge: Cambridge University Press, 103-139; Fleck, L. (2009). *Just Caring: Health Care Rationing and Democratic Deliberation*. Oxford: Oxford University Press.

End Notes for Science of Synthetic Biology

[1] Jackson, D.A., Symons, R.H., and P. Berg. (1972). Biochemical method for inserting new genetic information into DNA of simian virus 40: Circular SV40 DNA molecules containing lambda phage genes and the galactose operon of escherichia coli. *Proceedings of the National Academy of Sciences* 69(10):2904–2909.

[2] Jaenisch, R., and B. Mintz. (1974). Simian virus 40 DNA sequences in DNA of healthy adult mice derived from preimplantation blastocysts injected with viral DNA. *Proceedings of the National Academy of Sciences* 71(4):1250–1254.

[3] Twyman, R. (2003). *Transgenic Mice*. Available at: http://genome.wellcome.ac.uk/doc_WTD 021044.html.

[4] Berg, P. (2004). Asilomar and recombinant DNA. Available at: http://nobelprize.org/nobel_ prizes/chemistry/laureates/1980/berg-article.html.

[5] Fredrickson, D.S. (2001). *The Recombinant DNA Controversy: A Memoir*. Washington, D.C.: American Society for Microbiology Press.

[6] Goeddel, D., et al. (1979). Expression in *Escherichia coli* of chemically synthesized genes for human insulin. *Proceedings of the National Academy of Sciences* 76(1):106-110.

[7] Junod, S.W. (2007). Celebrating a milestone: FDA's approval of the first genetically-engineered product. *Selections from FDLI Update Series on FDA History*. Available at: http:// tinyurl.com/ybdj3sn.

[8] Gregory, T.R. (2010). *Animal Genome Size Database*. Available at: www.genomesize.com; Gibson, D.G., et al. (2010). Creation of a bacterial cell controlled by a chemically synthesized genome. *Science* 329(5987):52-56.

[9] Endy, D., Terman Fellow and Assistant Professor of Bioengineering, Stanford University. (2010). Overview and Context of the Science and Technology of Synthetic Biology. Presentation to the Presidential Commission for the Study of Bioethical Issues, July 8, 2010. Available at: www.bioethics.gov/ transcripts/ synthetic-biology/070810/overview-and-contextof-the-science-and-technology.html.

[10] Terrill, D., Senior Vice President and General Counsel for International Legal and Regulatory Affairs Integrated DNA Technologies. (2010). Biosafety and Biosecurity. Presentation to the President's Commission for the

Study of Bioethical Issues, September 14, 2010. Available at: http://www.bioethics.gov/transcripts/synthetic-biology/091410/ biosecurity-and-biosafety.html.

[11] Gibson, D.G., et al., op cit.

[12] Petsko, G.A. (2010). Hand-made biology. *Genome Biology* 11:124.

[13] Aldrich, S.C., Newcomb, J., and R. Carlson. (2007). Genome synthesis and design futures: Implications for the U.S. Economy. Cambridge, MA: Bio-Era. Available at: www.bio-era.net/ reports/genome.html. Current prices for gene synthesis vary, but have been observed as low at $0.39 per base pair. *See, e.g.*, Syd Labs. *Gene synthesis service*. Available at: http://tinyurl. com/2amzqwl.

[14] Fritz, B.R., et al. (2010). Biology by design: From top to bottom and back. *Journal of Biomedicine and Biotechnology* 2010:232016.

[15] Nature Biotechnology Editors (2009). Unbottling the genes. *Nature Biotechnology* 27:1059.

[16] Gil, R., Silva, F.J., and P.J. Moya. (2004). Determination of the core of a minimal bacterial gene set. *Microbiology and Molecular Biology Reviews* 68(3):518–537; Luisi, P.L., Ferri F., and P. Stano. (2006). Approaches to semi-synthetic minimal cells: A review. *Naturwissenschaften* 93(1):1–13; Glass, J.I., et al. (2006). Essential genes of a minimal bacterium. *Proceedings of the National Academy of Sciences* 103(2):425-430.

[17] Bayer T.S., et al. (2009). Synthesis of methyl halides from biomass using engineered microbes. *Journal of the American Chemical Society* 131(18):6508–6515.

[18] National Science Advisory Board for Biosecurity (NSABB). (2010). *Addressing Biosecurity Concerns Related to Synthetic Biology*. Available at: http://oba.od.nih.gov/biosecurity/pdf/ NSABB%20SynBio%20DRAFT%20 Report-FINAL%20(2)_6-7-10.pdf.

[19] King, J. (2010). Slippery cellularities: Analysis. *Seed*. Available at: http://seedmagazine.com/content/article/slippery_cellularities/; Cronin, L., et al. (2006). The imitation game—a computational chemical approach to recognizing life. *Nature Biotechnology* 24:1203-1206.

[20] National Science Advisory Board for Biosecurity. (2010). *Addressing Biosecurity Concerns Related to Synthetic Biology*. op cit., page ii.

[21] The BioBricks Foundation. Available at: http://bbf.openwetware.org/.

[22] Martin, V.J., et al. (2003). Engineering a mevalonate pathway in *Escherichia coli* from production of terpenoids. *Nature Biotechnology* 21(7):796-802.

[23] Dellomonaco, C., Fava. F., and R. Gonzalez. (2010). The path to next generation biofuels: Successes and challenges in the era of synthetic biology. *Microbial Cell Factories*. 9(3). Available at: www.microbialcell factories.com/content/9/1/3.

[24] Endy, D., op cit.; Weiss, R., Associate Professor, Department of Biological Engineering and Department of Electrical Engineering and Computer Science, Massachusetts Institute of Technology. (2010). Synthetic Biology: What New Methods and Products are Being Developed? Presentation to the Presidential Commission for the Study of Bioethical Issues, September 13, 2010. Available at: www.bioethics.gov/transcripts/synthetic-biology/091310/ synthetic-biology-what-new-methods-and-products-are-being-developed.html.

[25] Endy, D., op cit.

[26] Brenner, S., Senior Distinguished Fellow of the Crick-Jacobs Center, The Salk Institute. (2010). Reflections for the Future. Presentation to the Presidential Commission for the Study of Bioethical Issues, September 14, 2010. Available at: www.bioethics.gov/ transcripts/synthetic-biology/091410/reflections-for-the-future.html.

[27] Szybalski, W. (1974). In Vivo and in vitro initiation of transcription. *In Control of Gene Expression*, Kohn, A., and A. Shatkay (eds.). New York: Plenum Press. Page 405; *see also* Szybalski, W., and A. Skalka. (1978). Nobel prizes and restriction enzymes. *Gene* 4(3):181– 182.

[28] Keasling, J. D. (2010). Manufacturing molecules through metabolic engineering. *Science* 330:1355-1358.

[29] European Commission Framework Programme 6 Reference Contract 15357 (NEST). (2005). *Synbiology: An Analysis of Synthetic Biology Research in Europe and North America*. Available at: www2.spi.pt/synbiology/ documents/SYNBIOLOGY_Literature_And_ Statistical_Review.pdf.

[30] De Vriend, H. (2009). *Constructing Life: Early social reflections on the emerging field of synthetic biology (Working Document 97).* The Hague: Rathenau Institute. Available at: www.synbiosafe.eu/uploads/pdf/ Rathenau%20Constructing%20Life%202006.pdf.

[31] Keasling, J.D., op cit.

[32] National Research Council. (2010). *Sequence-based Classification of Select Agents: A Brighter Line*. Washington, D.C.: The National Academies Press. Available at: www.nap.edu/ catalog. php?record_id=12970; National Science Advisory Board for Biosecurity, op cit.

[33] Collins, Jj., University Professor, William F. Warren Distinguished Professor, Professor of Biomedical Engineering, and Co-Director, Center for BioDynamics, Boston University. (2010). Synthetic Biology: What

New Methods and Products are Being Developed? Presentation to the Presidential Commission for the Study of Bioethical Issues, September 13, 2010. Available at: www.bioethics.gov/transcripts/synthetic-biology/091310/syntheticbiology-what-new-methods-and-products-are-being-developed.html.

[34] Relman, D.A. (2010). Consideration of Advances in Synthetic Biology in Relation to the NSABB Report *Addressing Biosecurity Concerns Related to Synthetic Biology*: (1) Emerging Scientific Advances and (2) Creation of a Bacterial Cell Controlled by a Chemically Synthesized Genome. Presentation to National Science Advisory Board for Biosecurity (NSABB), October 19, 2010. Available at: http://oba.od.nih.gov/oba/biosecurity/meetings/ Oct2010/Consideration_of_Advances_in_SynBio_Slides.pdf.

End Notes for Applications, Benefits, and Risks

[1] Sheridan, C. (2009). Making green. *Nature Biotechnology* 27:1074 -1076.

[2] Department of Health and Human Services. (2004). National Science Advisory Board for Biosecurity Charter, March 4, 2004; National Science Advisory Board for Biosecurity Charter—Revised, March 10, 2006. Available at: http://oba.od.nih.gov/biosecurity/PDF/ NSABB%202010%20Charter_Renewal%20.pdf.

[3] Beer, L.L., et al. (2009). Engineering algae for biohydrogen and biofuel production. *Current Opinion in Biotechnology* 20(3):264-271.

[4] Savage, D.F., Way J., and P.A. Silver. (2008). Defossiling fuel: How synthetic biology can transform biofuel production. *American Chemical Society Chemical Biology* 3(1):13-16.

[5] Rulkens, W. (2008). Sewage sludge as a biomass resource for the production of energy: Overview and assessment of the various options. *Energy Fuels* 22(1):9-15.

[6] Synthetic Genomics, Inc. *Hydrocarbon Recovery & Conversion*. Available at: www. syntheticgenomics.com/what/hydrocarbonrecovery.html.

[7] For more information please visit: http://www.amyrisbiotech.com/.

[8] BP. *DuPont and BP Disclose Advanced Biofuels Partnership Targeting Multiple Butanol Molecules*. Available at: www.bp.com/genericarticle.do?categoryId=9024973&conten tId=7046879.

[9] Gevo. *Gevo Produces Isobutanol, Hydrocarbons and Jet Fuel from Cellulosic Biomass: Technology Milestone Validates Gevo's Biocatalyst for Converting Cellulosic Sugars into Advanced Biofuels and Hydrocarbons*. Available at: www.gevo.com/pr-gevo-producesisobutanol-hydrocarbons-jet-fuel-from-cellulosic-biomass.php.

[10] For more information please visit: http://www.global-bioenergies.com/.

[11] For more information please visit: http://www.jbei.org/.

[12] For more information please visit: http://www.ls9.com/.

[13] Bullis, K. (2007). Forget ethanol. *MIT Technology Review*. Available at: www. technology review.com/energy/18443/.

[14] Atsumi, S., Hanai, T., and J.C. Liao. (2008). Non-fermentative pathways for synthesis of branched-chain higher alcohols as biofuels. *Nature* 451:86-90.

[15] ExxonMobil. *Algae biofuels*. Available at: www.exxonmobil.com/Corporate/energy_climate_con_vehicle_algae.aspx.

[16] Synthetic Genomics, Inc. *Synthetic Genomics and EMRE Fact Sheet*. Available at: www. syntheticgenomics.com/media/emrefact.html.

[17] Chisti, Y. (2007). Biodiesel from microalgae. *Biotechnology Advances* 25(3):294-306; Solix. *About Microalgae and Biofuels*. Available at: www.solixbiofuels.com/content/technology/ about-microalgae-biofuels.

[18] For more information please visit: http://www.aurorainc.com/.

[19] For more information please visit: http://www.jouleunlimited.com/about/overview.

[20] For more information please visit: http://www.solazyme.com/.

[21] For more information please visit: http://www.syntheticgenomics.com/.

[22] Demirbas, A. (2009). *Biohydrogen for Future Engine Fuel Demands: Green Energy and Technology Virtual Series*. New York and Dordrecht: Springer. Pages 105-161 at 134.

[23] Clomburg, J.M., and R. Gonzales. (2010). Biofuel production in *Escherichia coli:* the role of metabolic engineering and synthetic biology. *Applied Microbiology and Biotechnology* 86:419-434 at 424.

[24] Beer, L.L., et al., op cit.

[25] Zhang, Y.H., et al. (2007). High-yield hydrogen production from starch and water by a synthetic enzymatic pathway. *PLoS One (Public Library of Science)* 2(5):e456; Science and technology: Gassed up—Synthetic biology. *Economist*. 383(8530):104. May 24, 2007; Ajudua, C. (2006). Driving: Y. H. Percival Zhang's crazy idea of the year: 'Sugar cars'. *Esquire*. November 30, 2006.

[26] Demirbas, A., op cit., at 134.

[27] Eggers, J., et al. (2009). Is biofuel policy harming biodiversity in Europe? *Global Change Biology Bioenergy* 1(1):18-34.

[28] Snow, A., Professor, Department of Evolution, Ecology & Organismal Biology, Ohio State University. (2010). Benefits and Risks of Synthetic Biology. Presentation to the Presidential Commission for the Study of Bioethical Issues, July 8, 2010. Available at: www.bioethics.gov/ transcripts/synthetic-biology/070810/ benefits-and-risks-of-synthetic-biology.html.

[29] Callura, J.M., et al. (2010). Tracking, tuning, and terminating microbial physiology using synthetic riboregulators. *Proceedings of the National Academy of Sciences* 107(36):15898- 15903.

[30] Thomas, J., Programme Manager, ETC Group. (2010). Benefits and Risks of Synthetic Biology. Presentation to the President's Commission for the Study of Bioethical issues, July 8, 2010. Available at: http://www.bioethics.gov/transcripts/synthetic-biology/070810/ benefitsand-risks-of-synthetic-biology.html.

[31] Dale, V.H., et al. (2010). Biofuels: Implications for land use and biodiversity. *Ecological Society of America: Biofuels and Sustainability Report*. Available at: www.esa.org/ biofuelsreports/files/ESA%20Biofuels%20 Report_VH%20Dale%20et%20al.pdf.

[32] Tilman, D., Reich., P.B., and J.M. Knops. (2006). Biodiversity and ecosystem stability in a decade-long grassland experiment. *Nature* 441:629-632; Tilman, D., et al. (2001). Diversity and productivity in a long-term grassland experiment. *Science* 294(5543):843-845.

[33] *See* Goeddel, D.V., et al. (1979). Expression in *Escherichia coli* of chemically synthesized genes for human insulin. *Proceedings of the National Academy of Sciences* 76(1):106–110; Mast, E.E., and J.W. Ward (2008). Hepatitis B vaccines. In *Vaccines*, S. Plotkin, W. Orenstein, and P. Offit (eds.). Pages 205-42; Schiller, J.T., Frazer, I.H., and Lowy, D.R. (2008). Human papillomavirus vaccines. In *Vaccines*, S. Plotkin, W. Orenstein, and P. Offit (eds.). Pages 243-258.

[34] Rodemeyer, M. (2009). *New Life, Old Bottles: Regulating First-Generation Products of Synthetic Biology*. Washington, D.C.: Woodrow Wilson International Center for Scholars. Available at: www.synbioproject.org/ process/assets/files/6319/nano_synbio2_electronic_ final. pdf.

[35] Centers for Disease Control and Prevention. *Malaria*. Available at: www.cdc.gov/ MALARIA/.

[36] Martin V.J., et al. (2003). Engineering a mevalonate pathway in *Escherichia coli* for production of terpenoids. *Nat Biotechnol*. 21:796-802.

[37] Amyris. *Artemisinin—Anti-malarial Therapeutic*. Available at: www.amyrisbiotech.com/ markets/artemisinin.

[38] Specter, M. (2009). A life of its own: Where will synthetic biology lead us? *The New Yorker*. Sept. 28, 2009. Pages 56-65, 58.

[39] Synthetic Genomics. *Press Release: Synthetic Genomics Inc. and J. Craig Venter Institute Form New Company, Synthetic Genomics Vaccines Inc. (SGVI), to Develop Next Generation Vaccines*. October 7, 2010. Available at: www.syntheticgenomics.com/media/ press/100710. html.

[40] Ibid.

[41] Yang, Z., et al. (2010). Expanded genetic alphabets in the polymerase chain reaction. *Angewandte Chemie International Edition* 49(1);177-180.

[42] Benner, S.A. (2004). Understanding nucleic acids using synthetic chemistry. *Accounts of Chemical Research* 37(10):784-797.

[43] Feero, E.G., Guttmacher, A.E., and F.S. Collins. (2010). Genomic medicine: An updated primer. *New England Journal of Medicine* 362:2001-2011.

[44] Weiss, R., Associate Professor, Department of Biological Engineering and Department of Electrical Engineering and Computer Science, Massachusetts Institute of Technology. (2010). Synthetic Biology: What New Methods and Products are Being Developed? Presentation to the Presidential Commission for the Study of Bioethical Issues, September 13, 2010. Available at: www.bioethics.gov/transcripts/synthetic-biology/091310/synthetic-biology-whatnew-methods-and-products-are-being-developed.html.

[45] Anderson, J.C., et al. (2006). Environmentally controlled invasion of cancer cells by engineered bacteria. *Journal of Molecular Biology* 355(4):619-627.

[46] Church, G.M., Professor of Genetics, Harvard Medical School. (2010). Applications of Synthetic Biology. Presentation to the Presidential Commission for the Study of Bioethical Issues, July 8, 2010. Available at: www.bioethics.gov/transcripts/synthetic-biology/070810/ applications-of-synthetic-biology.html.

[47] The Joint Bioenergy Institute. *About JBEI*. Available at: www.jbei.org; Synthetic Genomics, Inc.. *Agricultural Products*. Available at: www.syntheticgenomics.com/what/agriculture. html.

[48] Ibid.

[49] *See generally Diamond v. Chakrabarty*, 447 U.S. 303 (1980).

[50] Hazen, T.C., et al. (2010). Deep-sea oil plume enriches indigenous oil-degrading bacteria. *Science* 330(6001):204-208.

[51] For more information please visit: http://www.etcgroup.org/en.

[52] Scrinis, G. and Lyons, K., Nanotechnology and the Techno-Corporate Agri-Food Paradigm, in Lawrence, G., Lyons, K. and T. Wallington (eds.), *Food Security, Nutrition and Sustainability: New Challenges, Future Options,* Earthscan: London (2010).

[53] *See generally* de Lorenzo, V. (2008). Systems biology approaches to bioremediation. *Current Opinion in Biotechnology* 19(6): 579-589; Kirby, J.R. (2010). Synthetic biology: Designer bacteria degrades toxin. *Nature Chemical Biology* 6: 398-399.

[54] Cameotra, S.S., and R.S. Makkar. (2004). Recent applications of biosurfactants as biological and immunological molecules. *Current Opinion in Microbiology* 7(3):262-266.

[55] Snow, A.A., et al. (2005). Genetically engineered organisms and the environment: current status and recommendations. *Ecological Applications* 15:377-404.

[56] Norton, B.G., Distinguished Professor of Philosophy, School of Public Policy, Georgia Institute of Technology. (2010). Social Responsibility, Risk Assessment, and Ethics. Presentation to the Presidential Commission for the Study of Bioethical Issues, September 13, 2010. Available at: www.bioethics.gov/transcripts/synthetic-biology/091310/socialresponsibility-risk-assessment-and-ethics.html.

[57] Collins, Jj., University Professor, William F. Warren Distinguished Professor, Professor of Biomedical Engineering, and Co-Director, Center for BioDynamics, Boston University. (2010). Synthetic Biology: What New Methods and Products are Being Developed? Presentation to the Presidential Commission for the Study of Bioethical Issues, September 13, 2010. Available at: www.bioethics.gov/transcripts/synthetic-biology/091310/syntheticbiology-what-new-methods-and-products-are-being-developed.html

[58] NSABB. (2010). *Addressing Biosecurity Concerns Related to Synthetic Biology*. Available at: http://oba.od.nih.gov/biosecurity/pdf/NSABB%20SynBio%20DRAFT%20ReportFINAL%20(2)_6-7-10.pdf.

[59] Kodumal, Sj., et al. (2004). Total synthesis of long DNA sequences: Synthesis of a contiguous 32-kb polyketide synthase gene cluster. *Proceedings of the National Academy of Sciences* 101(44):15573-15578; Cello, J., Paul, A.V., and E. Wimmer. (2002). Chemical synthesis of poliovirus cDNA: Generation of infectious virus in the absence of natural template. *Science* 297(5583):1016-1018; Gibson, D.G., et al. (2008). Complete chemical synthesis, assembly, and cloning of a *Mycoplasma genitalium* genome. *Science* 319(5867):1215-1220; Tumpey, T.M., et al. (2005). Characterization of the reconstructed 1918 Spanish influenza pandemic virus. *Science* 310(5745):77-80.

[60] National Center for Biotechnology Information. *GenBank overview*. Available at: www.ncbi. nlm.nih. gov/genbank/.

[61] NSABB. (2006). *Addressing Biosecurity Concerns Related to the Synthesis of Select Agents*. Available at: http://oba.od.nih.gov/biosecurity/pdf/Final_NSABB_Report_on_Synthetic_ Genomics.pdf; NSABB, op cit. (2010).

[62] 42 C.F.R. Part 73.

[63] Department of Health and Human Services. (2010). *Screening Framework Guidance for Providers of Synthetic Double Stranded DNA*. Available at: www.phe.gov/Preparedness/ legal/guidance/syndna/Pages/default.aspx.

[64] *See, e.g.*, National Research Council, Board on Life Sciences. (2010). *Sequence-Based Classification of Select Agents: A Brighter Line*. Washington, D.C.: National Academies Press.

End Notes for Oversight

[1] Patterson, A., Acting Associate Director for Science Policy, NIH. (2010). Federal Oversight of Synthetic Biology Research. Presentation to the President's Commission for the Study of Bioethical Issues, July 9, 2010. Available at: www.bioethics.gov/documents/synthetic-biology/ Federal-Oversight-of-Synthetic-Biology-Research.ppt.

[2] *Coordinated Framework for Regulation of Biotechnology*, 51 Fed. Reg. 23302 (June 26, 1986).

[3] Patterson, A., op cit.

[4] Office of Biotechnology Activities. *About Recombinant DNA Advisory Committee (RAC)*. Available at: http://oba. od.nih.gov/rdna_rac/rac_about.html.

[5] DHHS/NIH. (2009). *NIH Guidelines for Recombinant DNA Molecules*. Available at: http:// oba.od.nih. gov/ rdna/nih_guidelines_oba.html.

[6] *Recombinant DNA Research: Proposed Actions Under the NIH Guidelines for Research Involving Recombinant DNA Molecules (NIH Guidelines)*, 74 Fed. Reg. 9411-9421 (Mar. 4, 2009).

[7] DHHS. (2009). *Biosafety in Microbiological and Biomedical Laboratories, 5th Edition*. Available at: www.cdc.gov/biosafety/publications/bmbl5/BMBL.pdf.

[8] APHIS/CDC. (2010). *Select Agents and Toxins*. Available at: www.selectagents.gov/ Select%20Agents%20 and%20Toxins.html.

[9] 42 C.F.R. § 73.3; 9 C.F.R. § 121.3; 7 C.F.R. § 331.3.

[10] APHIS/CDC. (2010). *Security Risk Assessments: Overview*. Available at: www.selectagents. gov/sra.html.

[11] Patterson, A., op cit.

[12] Rodemeyer, M., Lecturer, Department of Science, Technology and Society, School of Engineering and Applied Science, University of Virginia. (2010). Risks and Regulation of Products of Synthetic Biology Products. Presentation to the President's Commission for the Study of Bioethical Issues, July 9, 2010. Available at: www.bioethics.gov/transcripts/synthetic-biology/070910/federal-oversight-of-synthetic-biology.html; Rodemeyer, M. (2009). *New Life, Old Bottles: Regulating First-Generation Products of Synthetic Biology*. Report for the Woodrow Wilson International Center for Scholars Synthetic Biology Project/ Syn Bio 2. Page 40. Available at: www.synbioproject.org/library/ publications/ archive/ synbio2/.

[13] Ibid.

[14] Patterson, A., op cit. For example, in 2002 virologist Eckard Wimmer announced that his team had created live poliovirus "from scratch" using DNA they ordered by mail, and a viral genome map on the internet. Ball, P. (2004) Starting from scratch. *Nature* 431:624-626.

[15] *Screening Framework Guidance for Providers of Synthetic Double-Stranded DNA*, 75 Fed. Reg. 62820 (Oct. 13, 2010).

[16] APHIS/CDC. *Applicability of the Select Agent Regulations to Issues of Synthetic Genomics*. Available at: www.selectagents.gov/SyntheticGenomics.html.

[17] 7 C.F.R. Part 331; 9 C.F.R. Part 121; 42 C.F.R. Part 73.

[18] 7 U.S.C. § 8411.

[19] 7 C.F.R. § 331.3(c); 9 C.F.R. §§ 121.3(c), 121.4(c); 42 C.F.R. §§ 73.3(c), 73.4(c).

[20] APHIS/CDC. *Applicability of the Select Agent Regulations to Issues of Synthetic Genomics*, op cit.

[21] 7 C.F.R. § 331.9; 9 C.F.R. § 121.9; 42 C.F.R. § 73.9.

[22] 7 C.F.R. § 331.7; 9 C.F.R. § 121.7; 42 C.F.R. § 73.7.

[23] 7 C.F.R. § 331.17; 9 C.F.R. § 121.17; 42 C.F.R. § 73.17.

[24] 7 C.F.R. § 331.19; 9 C.F.R. § 121.19; 42 C.F.R. § 73.19.

[25] 7 C.F.R. § 331.18; 9 C.F.R. § 121.18; 42 C.F.R. § 73.18.

[26] APHIS/CDC. *Applicability of the Select Agent Regulations to Issues of Synthetic Genomics*, op cit.

[27] NSABB. (2006). *Addressing Biosecurity Concerns Relating to the Synthesis of Select Agents*. Available at: http://oba.od.nih.gov/biosecurity/pdf/Final_NSABB_Report_on_Synthetic_ Genomics.pdf.

[28] You, E.H., Supervisory Special Agent, Bioterrorism Program, Countermeasures Unit 1, FBI Weapons of Mass Destruction Directorate. (2010). FBI Perspective: Addressing Synthetic Biology and Biosecurity. Presentation to the President's Commission for the Study of Bioethical Issues, July 9, 2010. Available at: www.bioethics.gov/transcripts/syntheticbiology/070910/federal-oversight-of-synthetic-biology.html.

[29] Bureau of Industry and Security. *Policies and Regulations*. Available at: www.bis.doc.gov/policies andregulations/index.htm.

[30] Excepted are items of national security or foreign policy significance that are exclusively controlled by enumerated other agencies. 15 C.F.R. § 734.3.

[31] 15 C.F.R. § 730.3.

[32] 15 C.F.R. § 738.1.

[33] 15 C.F.R. § 738.2(a).

[34] Bureau of Industry and Security. *Lists to Check*. Available at: www.bis.doc.gov/complianceand enforcement/liststocheck.htm.

[35] 49 U.S.C. § 5103.

[36] 49 C.F.R. § 171.8.

[37] 49 C.F.R. § 173.24.

[38] 49 C.F.R. Part 172, Subpart E.

[39] 49 C.F.R. Part 172, Subpart H.

[40] 49 C.F.R. §§ 107.309-107.310.

[41] 49 C.F.R. § 107.329.

[42] 49 C.F.R. § 107.333.

[43] 49 C.F.R. §§ 174-177.

[44] Screening Framework Guidance for Providers of Synthetic Double-Stranded DNA (2010), op cit., at 62820.

[45] Ibid.

[46] *Screening Framework Guidance for Synthetic Double-Stranded DNA Providers*, 74 Fed. Reg. 62319, 62319 (Nov. 27, 2009).

[47] Ibid.

[48] *Screening Framework Guidance for Providers of Synthetic Double-Stranded DNA (2010)*, op cit.

[49] DHHS/NIH, op cit. at Sec. 1-C.

[50] Fauci, A.S., Director, National Institute of Allergy and Infectious Diseases. (2010). Advances in Synthetic Biology: Significance and Implications. Testimony before the Committee on Energy and Commerce of the U.S. House of Representatives, May 27, 2010. Available at: http://energycommerce.house.gov/ documents/20100527/Fauci.Testimony.05.27.2010.pdf.

[51] DHHS, op cit.

[52] Ibid.

[53] Office of Biotechnology Activities. (2009). *Frequently Asked Questions (FAQs) of Interest to IBCs*. Available at: http://oba.od.nih.gov/oba/ibc/FAQs/IBC_Frequently_Asked_ Questions7.24.09.pdf.

[54] DHHS, op cit.

[55] *NIH Guidelines*, op cit.

[56] Venter, J.C., Founder and President, JCVI. (2010). Applications of Synthetic Biology. Presentation to the President's Commission for the Study of Bioethical Issues, July 8, 2010. Available at: http://www.bioethics. gov/transcripts/synthetic-biology/070810/applications-ofsynthetic-biology.html.

[57] DHHS, op cit.

[58] Ibid.

[59] Congressional Research Service. (2009). *Oversight of High-Containment Biological Laboratories: Issues for Congress*. Pages 8-9. Available at: www.fas.org/sgp/crs/ terror/R40418. pdf.

[60] 29 U.S.C. § 651.

[61] OSHA. *State Occupational Safety and Health Plans*. Available at: http://www.osha.gov/dcsp/ osp/index.html.

[62] 29 U.S.C. § 654.

[63] 29 C.F.R. Part 1910.

[64] 29 C.F.R. § 1910.120.

[65] 29 C.F.R § 1910.1200.

[66] 29 C.F.R. § 1910.1030(a)-(b).

[67] 29 C.F.R. § 1910.1030(c)-(f).

[68] 40 C.F.R. Part 721.

[69] Willis, J., Director, Chemical Control Division, Office of Pollution Prevention and Toxics, EPA. (2010). Comments submitted to the Presidential Commission for the Study of Bioethical Issues, July 19, 2010.

[70] 40 C.F.R. § 725; EPA. (2010). *Microbial Products of Biotechnology: Summary of Regulations Under the Toxic Substances Control Act (Fact Sheet)*. Available at: www.epa.gov/biotech_ rule/pubs/fs-001.htm.

[71] 15 U.S.C. § 2604; 40 C.F.R. Part 725.

[72] 40 C.F.R. Part 725.

[73] 40 C.F.R. § 725.250.

[74] 40 C.F.R. Part 725.

[75] 15 U.S.C. § 2607(a)(2).

[76] Rodemeyer, M. (2009), op cit.

[77] 15 U.S.C. § 2607(e).

[78] Rodemeyer, M. (2009), op cit.

[79] 21 U.S.C. § 355; 21 C.F.R. Part 314; 21 U.S.C § 360(k); 21 C.F.R. Part 807, Subpart E; 21 U.S.C. § 360e; 21 C.F.R Part 814; FDA. (2010). *How Drugs Are Developed and Approved*. Available at: www.fda.gov/Drugs/DevelopmentApprovalProcess/ HowDrugsare Developed andApproved/default.htm; FDA. (2009). *Overview of Device Regulation*. Available at: www.fda.gov/MedicalDevices/Device RegulationandGuidance/ Overview/ default.htm.

[80] FDA. (2009). *FDA News Release: FDA Approves Orphan Drug ATryn to Treat Rare Clotting Disorder*. Available at: www.fda.gov/NewsEvents/Newsroom/PressAnnouncements/2009/ ucml09074.htm; FDA. (2007). *Guidance for Industry and FDA Staff: Class II Special Controls Guidance Document: Absorbable Poly(hydroxybutyrate) Surgical Suture Produced by Recombinant DNA Technology*. Available at: www.fda.gov/downloads/MedicalDevices/DeviceRegulationandGuidance/Guidance Documents/ucm071352. pdf.

[81] Junod, S.W. (2007). *Celebrating a Milestone: FDA's Approval of First Genetically-Engineered Product.* Available at: www.fda.gov/AboutFDA/WhatWeDo/History/ProductRegulation/ SelectionsFromFDLIUpdateSeriesonFDAHistory/ucm081964.htm.

[82] FDA. (1997). *Guidance for Industry: ICH S6 Preclinical Safety Evaluation of Biotechnology-Derived Pharmaceuticals.* Available at: www.fda.gov/downloads/Drugs/ Guidance complianceRegulatoryInformation/ Guidances/ucm074957.pdf; FDA. (2009). *Draft Addendum to ICH S6: Preclinical Safety Evaluation of Biotechnology-Derived Pharmaceuticals S6(R1).* Available at: www.fda.gov/downloads/Drugs/Guidance ComplianceRegulatoryInformation/Guidances/UCM194490.pdf; FDA. (2002). Draft Guidance for Industry: Drugs, Biologics, and Medical Devices Derived from Bioengineered Plants for Use in Humans and Animals. Available at: www.fda.gov/ downloads/Drugs/GuidanceComplianceRegulatoryInformation/ Guidances/ucm 124811. pdf; FDA. (2009). *Guidance for Industry: Regulation of Genetically Engineered Animals Containing Heritable Recombinant DNA Constructs.* Available at: www.fda.gov/downloads/ AnimalVeterinary/ GuidanceComplianceEnforcement/Guidancefor Industry/UCM113903. pdf.

[83] FDA. (2009). *Animal and Veterinary General Q & A: What Is Genetic Engineering?* Available at: www.fda.gov/AnimalVeterinary/DevelopmentApprovalProcess/GeneticEngineering/GeneticallyEngineered Animals/ucm113605.htm.

[84] 21 C.F.R. Part 312; 21 C.F.R Part 314; 21 C.F.R. § 314.80; 21 C.F.R. § 314.81.

[85] 21 C.F.R. Part 58; 21 C.F.R. Parts 210 and 211; 21 C.F.R. Parts 201-203.

[86] 21 C.F.R Part 50; 21 C.F.R. Part 56; 21 C.F.R. § 312.32; 21 C.F.R. § 812.150.

[87] FDA. (2009). *Draft Guidance for Industry: Format and Content of Proposed Risk Evaluation and Mitigation Strategies (REMS), REMS Assessments, and Proposed REMS Modifications.* Available at: www.fda.gov/downloads/Drugs/GuidanceComplianceRegulatoryInformation/ Guidances/UCM184128.pdf.

[88] 21 C.F.R. § 314.150; 21 C.F.R. § 814.46; 21 C.F.R. Part 7, Subpart C; 21 U.S.C. § 332; 21 U.S.C. § 334.

[89] 21 U.S.C. § 333.

[90] FDA. (1985). *Draft: Points to Consider in the Production and Testing of New Drugs and Biologicals Produced by Recombinant DNA Technology.* Available at: www.fda. gov/downloads/BiologicsBloodVaccines/ GuidanceComplianceRegulatoryInformation/ OtherRecommendationsforManufacturers/UCM062750.pdf.

[91] FDA. (2010). *Veterinary Medicine Advisory Committee Briefing Packet: AquAdvantage Salmon.* Page 1. Available at: www.fda.gov/downloads/AdvisoryCommittees/ Committees MeetingMaterials/Veterinary MedicineAdvisoryCommittee/UCM224762.pdf.

[92] FDA. (2010). *Regulation of Genetically Engineered Animals.* Available at: www.fda. gov/downloads/For Consumers/ConsumerUpdates/ucm048122.pdf; *Int'l Ctr. for Tech. Assessment v. Thompson*, 421 F. Supp. 2d 1 (D.D.C. 2006); *Int'l Ctr. for Tech. Assessment v. Leavitt*, 468 F. Supp. 2d 200 (D.D.C. 2007).

[93] 42 U.S.C. § 4321 *et seq.*; Council on Environmental Quality, Executive Office of the President. (2007). *A Citizen's Guide to the NEPA: Having Your Voice Heard.* Page 13. Available at: http://ceq.hss.doe. gov/nepa/Citizens_Guide_Dec07.pdf.

[94] Ibid, at 5, 11.

[95] Ibid, at 13-14.

[96] Ibid, at 6.

[97] Ibid, at 5.

[98] See generally, APHIS. (2007). *Biotechnology.* Available at: www.aphis.usda.gov/ biotechnology/index.shtml.

[99] 7 C.F.R. § 340.1.

[100] 7 U.S.C. §§ 7701-7736.

[101] 7 U.S.C. § 7702(14).

[102] 7 C.F.R. Part 340.

[103] APHIS. (2007). *Permits, Notifications, and Petitions.* Available at: www.aphis.usda.gov/ biotechnology/submissions.shtml; 7 C.F.R. § 340.3; 7 C.F.R. § 340.4.

[104] 7 C.F.R. § 340.4(g).

[105] 7 C.F.R. § 340.6.

[106] APHIS. (2010). *Petitions.* Available at: www.aphis.usda.gov/biotechnology/petitions.shtml.

[107] 9 C.F.R. § 102.3(b).

[108] 9 C.F.R. § 102.4.

[109] 9 C.F.R. § 103.2.

[110] 9 C.F.R. § 103.3.

[111] 9 C.F.R. § 104.1(a).

[112] 9 C.F.R. § 104.2(a).

[113] 9 C.F.R. § 104.2.

[114] 9 C.F.R. Part 115.
[115] 9 C.F.R. Part 116.
[116] 7 U.S.C. § 136 *et seq*.
[117] 7 U.S.C. § 136a(c)(5); 7 U.S.C. § 136(bb). *See also* EPA. (2010). *Regulating Pesticides*. Available at: www.epa.gov/pesticides/regulating/index.htm.
[118] EPA. (2010). *Evaluating New Pesticides and Uses*. Available at: www.epa.gov/pesticides/regulating/index.htm#eval.
[119] 40 C.F.R. Parts 152, 158, and 161; EPA. (2010). *Pesticide Registration Program*. Available at: www.epa.gov/pesticides/factsheets/registration.htm#process.
[120] 21 U.S.C. § 346a; EPA. (2010). *Pesticide Tolerances*. Available at: www.epa.gov/pesticides/regulating/tolerances.htm.
[121] 40 C.F.R. § 172.3(b)(1).
[122] 7 U.S.C. § 136c; 40 C.F.R. Part 172, Subpart A; 40 C.F.R. § 172.45.
[123] American Association of Pesticide Control Officials. Available at: http://aapco.ceris.purdue. edu/index.html.
[124] 21 U.S.C. § 348; International Food Information Council and FDA. (2010). *Food Ingredients and Colors*. Available at: www.fda.gov/Food/FoodIngredientsPackaging/ ucm094211.htm; FDA. (1992). *Statement of Policy: Foods Derived from New Plant Varieties: Guidance to Industry for Foods Derived from New Plant Varieties*, 57 Fed. Reg. 22984 (May 29, 1992).
[125] 21 U.S.C § 334; 21 U.S.C. § 342.
[126] 21 C.F.R. § 173.170; 21 C.F.R. § 573.130; FDA. (1994). *Agency Summary Memorandum Re: Consultation with Calgene, Inc., Concerning FLAVR SAVRTM Tomatoes*. Available at: www. fda.gov/Food/Biotechnology/Submissions/ucm225043.htm#out38.
[127] Maryanski, J.H., Biotechnology Coordinator, Center for Food Safety and Applied Nutrition, FDA. (1999). *Genetically Engineered Foods*. Statement to the Subcommittee on Basic Research, House Committee on Science, October 19, 1999. Available at: www.fda.gov/ newsevents/testimony/ucm115032.htm.
[128] 21 U.S.C. § 343(a).
[129] FDA. (2010). *Background Document: Public Hearing on the Labeling of Food Made from the AquAdvantage Salmon*, Page 4. Available at: www.fda.gov/downloads/ Food/Labeling Nutrition/FoodLabelingGuidance RegulatoryInformation/TopicSpecificLabeling Information/UCM223913.pdf; FDA Statement of Policy, op cit.; FDA. (2001). *Draft Guidance for Industry: Voluntary Labeling Indicating Whether Foods Have or Have Not Been Developed Using Bioengineering*. Available at: www.fda.gov/food/ guidance complianceregulatoryinformation/guidancedocuments/foodlabelingnutrition/ ucm059098.htm.
[130] FDA Background Document, op cit.
[131] *Alliance for Bio-Integrity v. Shalala*, 116 F. Supp. 2d 166, 178-179 (D.D.C. 2000).
[132] 40 C.F.R. Part 68.
[133] 40 C.F.R. Part 112; 40 C.F.R. Part 110.
[134] 40 C.F.R. Part 68; 40 C.F.R. Part 112; 40 C.F.R. Part 110; 40 C.F.R Part 355.

End Notes for Analysis and Recommendations

[1] Although current levels of public and private investment in synthetic biology are not available, according to the Woodrow Wilson Center, the U.S. Government has spent approximately $430 million on research related to synthetic biology since 2005. By comparison, the European Union and three individual European countries—The Netherlands, U.K., and Germany—spent approximately $160 million during that same period. See: Woodrow Wilson International Center for Scholars. (2010). *Trends in Synthetic Biology Research Funding in the United States and Europe, June 2010, Research Brief 1*. Available at: http://www.synbioproject.org/library/publications/archive/researchfunding/. A few companies' combined spending already far exceeds those public investments. (*See, e.g.*, Dimond, P. (2010). Analysis & Insight: Synthetic biology—the devil is in the financial details. *Genetic Engineering and Biotechnology News* (July 13, 2010). Available at: http://www.genengnews.com/analysis-and-insight/synthetic-biology-the-devil-is-in-the-financialdetails/77899331/.)
[2] Roll Back Malaria. (2010). *Key malaria facts*. Available at: http://www.rbm.who.int/ keyfacts.html.
[3] Rejeski, D., Director, Science and Technology Innovation Program, Woodrow Wilson International Center for Scholars. (2010). Synthetic Biology, the Public, and the Media. Presentation to the Presidential Commission for the Study of Bioethical Issues, July 9, 2010. http://bioethics.gov/transcripts/synthetic-biology/070910/ethics-of-synthetic-biology.html; Snow, A. Professor, Department of Evolution, Ecology &

Organismal Biology, Ohio State University. (2010). Risks of Environmental Releases of Synthetic GEOs. Presentation to the Presidential Commission for the Study of Bioethical Issues, July 9, 2010. http://bioethics.gov/ transcripts/synthetic-biology/070810/benefits-and-risks-of-synthetic-biology.html.

4 *See, e.g.*, Cho, M., et al. (1999). Ethical considerations in synthesizing a minimal genome. Science 286(5447):2087-2090; The Alfred P. Sloan Foundation. Basic Research: Synthetic Biology. Available at: http://www.sloan.org/program/38.

5 NIH. (2010). *The NIH Common Fund—High Risk Research*. Available at: http:// nihroadmap. nih.gov/highrisk/.

6 The National Institutes of Health Reform Act of 2006. Pub. L. 109-482, 120 Stat. 3677 (2007).

7 The Bill and Melinda Gates Foundation. *Grand Challenges in Global Health*. Available at: http://www.grandchallenges.org/explorations/Pages/introduction.aspx.

8 Kohanski, M., Dwyer, D., and J. Collins. (2010). How antibiotics kill bacteria: From targets to networks. *Nature Reviews Microbiology* 8 (June 2010):423-435; Pollack, A. (2010). Antibiotics Research Subsidies Weighed by U.S. *New York Times*. November 5, 2010.

9 Erickson, B., on behalf of the Biotechnology Industry Organization. (2010). Comments submitted to the Presidential Commission for the Study of Bioethical Issues, October 1, 2010.

10 Eisenberg, R.S. (2010). Patents and data-sharing in public science. *Industrial and Corporate Change* 15(6):1013-1031.

11 *See* Secretary's Advisory Committee on Genetics, Health, and Society. (2010). *Revised Draft Report on Gene Patents and Licensing Practices and Their Impact on Patient Access to Genetic Tests*. Available at: http:// oba.od.nih.gov/oba/SACGHS/SACGHS%20Patents%20 Report%20Approved%202-5-20010.pdf; see also, National Research Council (2006). *Reaping the Benefits of Genomic and Proteomic Research*. Washington, D.C.: National Academies Press.

12 *Ass'n for Molecular Pathology, et al. v. U.S. Patent and Trademark Office, et al.*, 702 F. Supp. 2d 181 (S.D.N.Y. 2010), amended by 2010 U.S. Dist. LEXIS 35418 (S.D.N.Y. 2010), *appeal filed*, No. 2010-1406 (Fed. Cir. filed June 16, 2010).

13 Henkel, J., and S.M. Maurer. (2009). Parts, property and sharing. *Nature Biotechnology* 27(12):1095-1098; See also, Rai, A.K., Elvin R. Latty Professor of Law, Duke University School of Law, Center for Public Genomics, Institute for Genome Sciences and Policy, Duke University. (2010). Knowledge Sharing, Innovation and Translating Research for the Public Good. Presentation to the President's Commission for the Study of Bioethical Issues, September 13, 2010. Available at: http://www.bioethics.gov/transcripts/synthetic-biology/ 091310/knowledge-sharing-innovation-and-translating-research-for-the-public-good. html.

14 SACGHS, op cit. See also Brief for the United States as Amicus Curiae in Support of Neither Party, *Ass'n for Molecular Pathology, et al. v. U.S. Patent and Trademark Office, et al.*, 702 F. Supp. 2d 181 (S.D.N.Y. 2010).

15 Balmer, A., and P. Martin. (2008). *Synthetic Biology: Social and Ethical Challenges* (An independent review commissioned by the Biotechnology and Biological Sciences Research Council (BBSRC)). Available at: http://www.bbsrc.ac.uk/organisation/policies/reviews/ scientific-areas/0806-synthetic-biology.aspx

16 ETC Group. (2007). Backgrounder: J. Craig Venter Institute's patent application on world's first human-made species. Available at: http://www.etcgroup.org/en/materials/publications. html?pub_id=631.

17 The BioBricks Foundation. Available at: http://biobricks.org/.

18 Principles and Guidelines for Recipients of NIH Research Grants and Contracts on Obtaining and Disseminating Biomedical Research Resources: Final Notice, 64 Fed. Reg. 72090-72096 (Dec. 23, 1999); NIH. (2003). NIH Data Sharing Policy and Implementation Guidance. Available at: http://grants.nih.gov/grants/policy/ data_sharing/data_sharing_ guidance.htm; Consolidated Appropriations Act of 2008, Pub. L. No 110-161, §218, 121 Stat. 1844 (2007).

19 The Howard Hughes Medical Institute. *Research Policies: Confidential Disclosure Agreements* (SC-370). Available at: http://www.hhmi.org/about/research/sc370.pdf; the Wellcome Trust. *Wellcome Trust Policy on Conflicts of Interest and Related Standards of Conduct*. Available at: http://www.wellcome.ac.uk/stellent/ groups/corporatesite/@legal/ documents/ web_document/ wtd002770.pdf.

20 International Committee of Medical Journal Editors (ICMJE) website. Toward More Uniform Conflict Disclosures: The Updated ICMJE Conflict of Interest Reporting Form Disclosure Policy. Available at: http://www.icmje.org/updated_coi.pdf; Food and Drug Administration Amendments Act of 2007, Pub. L. No 110-85,121 Stat. 823 (2007); 42 U.S.C. § 282(i).

21 Rai, A.K., op cit.; Stevens, A.J., Special Assistant to the Vice President of Research, Boston University. (2010). Knowledge Sharing, Innovation and Translating Research for the Public Good. Presentation to the Presidential Commission for the Study of Bioethical Issues, September 13, 2010. Available at: http://bioethics.gov/ transcripts/synthetic-biology/091310/ knowledge-sharing-innovation-and-translating-research-for-the-public-good.html.

[22] *Ass'n for Molecular Pathology*, op cit.

[23] For a more extensive discussion, see: Epstein, L.S. (1980). Decision-making and the temporal resolution of uncertainty. *International Economic Review* 21(2):269–283; Arrow, Kj. and A.C. Fisher. (1974). Environmental preservation, uncertainty and irreversibility. *Quarterly Journal of Economics* 88(2):312–319; Gollier, C., Bruno, J., and N. Treich. (2000). Scientific progress and irreversibility: An economic interpretation of the 'precautionary principle.' *Journal of Public Economics* 75(2):229–253. See also the Wingspread Consensus Statement at: http://www.sehn.org/wing.html.

[24] Commission of the European Communities. (2000). Communication for the Commission on the Precautionary Principle. Brussels 02.02.2000 COM(2000) 1. Available at: http:// ec.europa.eu/dgs/health_consumer/ library/pub/pub07_en.pdf.

[25] Parens, E., Johnston, J., and J. Moses. (2009). *Ethical Issues in Synthetic Biology: An Overview of the Debates*. Washington, D.C.: Woodrow Wilson International Center for Scholars. Available at: http://www.syn bioproject.org/process/assets/files/6334/synbio3.pdf.

[26] Ibid.

[27] Ibid.

[28] NSABB. (2010). *Addressing Biosecurity Concerns Related to Synthetic Biology*. Available at: http://oba.od.nih. gov/biosecurity/pdf/NSABB%20SynBio%20DRAFT%20ReportFINAL%20(2)_6-7-10.pdf. Page 6.

[29] Patterson, A., Acting Associate Director for Science Policy, National Institutes of Health. (2010). Federal Oversight of Synthetic Biology Research. Presentation to the President's Commission for the Study of Bioethical Issues, July 9, 2010. Available at: http://www. bioethics.gov/transcripts/synthetic-biology/070910/federal-oversight-of-synthetic-biology. html.

[30] It is noteworthy that the Recombinant DNA Advisory Committee has found that, in most cases, "research with synthetic nucleic acids presents biosafety risks that are comparable to recombinant DNA research and that certain work with synthetic nucleic acids in a non-replicating form may not require oversight under the *NIH Guidelines* (although other biosafety standards would apply)." *See* NSABB, op cit., p.9.

[31] DHHS, Screening Framework Guidance for Synthetic Double-Stranded DNA Providers. *Federal Register* 74(227):62319-62327. November 27, 2009.

[32] APHIS/CDC. Applicability of the Select Agent Regulations to Issues of Synthetic Genomics. Available at: http://www.selectagents.gov/syntheticgenomics.html.

[33] Patterson, A., op cit.

[34] Swiss Confederation. The Swiss Federal Ethics Committee on Non-Human Biotechnology. (May 2010). *Synthetic Biolog—Ethical Considerations*. Available at: http://www.ekah. admin.ch/fileadmin/ekah-dateien/ dokumentation/publikationen/e-Synthetische_Bio_ Broschuere.pdf; International Risk Governance Council. (2009). *Concept Note: Risk Governance of Synthetic Biology*. Geneva: International Risk Governance Council. Available at: http://www. irgc.org/IMG/pdf/IRGC_Concept_Note_Synthetic_Biology_ 191009_FINAL.pdf.

[35] See Tucker, J.B. (2010). Double-edged DNA: Preventing the misuse of gene synthesis. *Issues in Science and Technology* (Spring 2010).

[36] Keasling, J. (2010). Testimony to Committee on Energy and Commerce, U.S. House of Representatives, May 27, 2010. Available at: http://energycommerce.house.gov/ documents/20100527/Keasling. Testimony.05.27.2010. pdf.

[37] Synth-Ethics. *Ethical and regulatory challenges raised by synthetic biology*. Available at: http:// synthethics.eu/.

[38] Maurer, S., and L. Zoloth. (2007). Synthesizing biosecurity. *Bulletin of the Atomic Scientists* 63(6):16-18; Garfinkel, M.S., et al. (2008). *Synthetic Genomics: Options for Governance*. Rockville, Maryland: The J. Craig Venter Institute. Available at: http://www.jcvi.org/cms/ fileadmin/site/research/projects/synthetic-genomics-report/synthetic-genomics-report.pdf.

[39] Terrill, D., Senior Vice President and General Counsel for International and Legal Affairs, Integrated DNA Technologies, and Chief Operating Officer, International Consortium for Polynucleotide Synthesis. (2010). Biosecurity and Biosafety. Presentation to the Presidential Commission for the Study of Bioethical Issues, September 14, 2010; Wagner, R., Chief Executive Officer and Chief Scientific Officer, Geneart AG, and Professor and Chief, Molecular Microbiology and Gene Therapy, University of Regensburg Institute of Medical Microbiology and Hygiene. (2010). Biosecurity and Biosafety. Presentation to the Presidential Commission for the Study of Bioethical Issues, September 14, 2010. Available at: http://www.bioethics.gov/transcripts/synthetic-biology/091410/biosecurity-and-bio safety.html.

[40] Ibid.

[41] Church, G.M. (2009) Safeguarding biology. *Seed* (February 2009):85-86. Available at: http:// arep.med.harvard.edu/pdf/Church_Seed09.pdf .

[42] Callura, J.M., et al. (2010). Tracking, tuning, and terminating microbial physiology using synthetic riboregulators. *Proceedings of the National Academy of Sciences*. Published online before print. Available at: http://www.pnas.org/content/early/2010/08/09/ 1009747107.full. pdf+html.

[43] Snow, A.A. (2003). Genetic engineering: Unnatural selection. *Nature* 424(7 Aug 2003):619; Snow, A.A., et al. (2005). ESA Report: Genetically engineered organisms and the environment: Current status and recommendations. *Ecological Applications*, 15(2):377–404.

[44] Rodemeyer, M. (2009). *New Life, Old Bottles: Regulating First-Generation Products of Synthetic Biology*. Report for the Woodrow Wilson International Center for Scholars Synthetic Biology Project/Syn Bio 2. Available at: www.synbioproject.org/library/ publications/archive/synbio2/; Rodemeyer, M., Lecturer, Department of Science, Technology and Society, School of Engineering and Applied Science, University of Virginia. (2010). Federal Oversight of Synthetic Biology. Presentation to the Presidential Commission for the Study of Bioethical Issues, July 9, 2010. Available at: http://www.bioethics.gov/transcripts/synthetic-biology/ 070910/federal-oversight-of-synthe tic-biology. html.

[45] *See, e.g.*, Committee on Laboratory Security and Personnel Reliability Assurance Systems for Laboratories Conducting Research on Biological Select Agents and Toxins; National Research Council. (2009). *Responsible Research with Biological Select Agents and Toxins*. Washington, D.C.: National Academies Press.

[46] NIH. (2000). Notice OD-00-039: Required Education in the Protection of Human Research Participants. Available at: http://grants.nih.gov/grants/guide/notice-files/NOT-OD00-039.html

[47] 7 C.F.R § 331.15; 9 C.F.R § 121.15; 42 C.F.R § 73.15.

[48] DHHS/NIH. (2009). *NIH Guidelines for Recombinant DNA Molecules*. Sec. III. Available at: http://oba.od.nih.gov/rdna/nih_guidelines_oba.html.

[49] America COMPETES Act, Pub. L. No. 110-69, § 7009, 121 Stat. 572-718 (2007).

[50] NSABB, op cit.

[51] *See, e.g.*, Kaebnick, G. (2009). Should moral objections to synthetic biology affect public policy? *Nature Biotechnology* 27(12):1106-1108.

[52] Kaebnick, G. Research Scholar, The Hastings Center. (2010). Ethics of Synthetic Biology. Presentation to the Presidential Commission for the Study of Bioethical Issues, July 8, 2010. Available at: http://www.bioethics.gov/transcripts/synthetic-biology/070810/ethics-ofsynthetic-biology.html.

[53] Ibid.

[54] Caplan, A.L., Emmanuel and Robert Hart Director, Center for Bioethics, and Sidney D. Caplan Professor of Bioethics, School of Medicine, University of Pennsylvania. (2010). Philosophical and Theological Perspectives. Presentation to the Presidential Commission for the Study of Bioethical Issues, September 13, 2010. Available at: http://www. bioethics.gov/transcripts/synthetic-biology/091310/philosophical-and-theological-perspectives.html.

[55] Boldt, J., and O. Muller. (2008). Newtons of the leaves of grass. *Nature Biotechnology* 26(4):387-389.

[56] Ibid.

[57] Douglas, T., and J. Savulescu. (2010). Synthetic biology and the ethics of knowledge. *Journal of Medical Ethics* 36(11):687-693.

[58] Ibid.

[59] ETC Group. (2010). *The New Biomassters: Synthetic Biology and the Next Assault on Biodiversity and Livelihoods*. Available at: http://www.etcgroup.org/upload/publication/ pdf_file/biomassters.pdf.

[60] Buchanan, A., James B. Duke Professor of Philosophy, Duke University; Investigator, Institute for Genome Sciences and Policy, Duke University; Distinguished Research Associate, Uehiro Centre for Practical Ethics, University of Oxford. (2010). Ethics of Synthetic Biology. Presentation to the Presidential Commission for the Study of Bioethical Issues, July 8, 2010. Available at: http://www.bioethics.gov/transcripts/synthetic-biology/070810/ ethics-of-synthetic-biology.html.

[61] Preston, C. (2008). Synthetic biology: Drawing a line in Darwin's sand. *Environmental Values* 17:23-39.

[62] ETC Group (2010), op cit.

[63] Ibid.

[64] BBC Monitoring Europe. *Vatican Dismisses Synthetic Cell's Life-Giving Dimension, Lauds Science Research*. May 25, 2010. Translation of Gian Guido Vecchi. *Cautela in Vaticano 'È un ottimo motore ma non è la vita.'* Available at: http://archiviostorico.corriere.it/2010/ maggio/22/Cautela_Vaticano_ottimo_motore_non_co_9_100522032.shtml.

[65] Wheeler, S.E., Martha Ashby Carr Professor of Christian Ethics, Wesley Theological Seminary. (2010). Contributions of Christian Thought to Assessments of Synthetic Biology. Presentation to the Presidential Commission for the Study of Bioethical Issues, September 13, 2010. Available at:

http://www.bioethics.gov/transcripts/synthetic-biology/091310/philosophical-and-theological-perspectives.html.

[66] Ibid.

[67] Buchanan, A., op cit.; See also Lustig, B.A., Brody, B.A., and G.P. McKenny (eds.). (2008). *Altering Nature. Vol. I: Concepts of 'Nature' and 'The Natural' in Biotechnology Debates*. New York: Springer Publishing.

[68] American Association of University Professors. (1940). *Statement of Principles on Academic Freedom and Tenure*. Available at: http://www.aaup.org/AAUP/pubsres/policydocs/ contents/1940statement.htm. Consider also the discussion by Supreme Court Chief Justice Warren, in *Sweezey v. New Hampshire*, 354 U.S. 234 (1957), defending the academic freedom of a faculty member and recognizing the importance of new knowledge to civilization: "No field of education is so thoroughly comprehended by man that new discoveries cannot yet be made...Scholarship cannot flourish in an atmosphere of suspicion and distrust. Teachers and students must always remain free to inquire, to study and to evaluate, to gain new maturity and understanding; otherwise our civilization will stagnate and die." *Sweezey v. New Hampshire*, 354 U.S. 234, 250 (1957).

[69] *See* You, E.H., Supervisory Special Agent, Federal Bureau of Investigation. (2010). FBI Perspective: Addressing Synthetic Biology and Biosecurity. Federal Oversight of Synthetic Biology. Presentation to the Presidential Commission for the Study of Bioethical Issues, July 9, 2010. Available at: http://www.bioethics.gov/ transcripts/synthetic-biology/070910/ federaloversight-of-synthetic-biology.html.

[70] Berg, P., et al. (1975). Summary statement of the Asilomar conference on Recombinant DNA molecules. *Proceedings of the National Academy of Sciences*. 72(6);1981-1984 at p. 1981.

[71] *See, e.g.*, DHHS. (2009). *Biosafety in Microbiological and Biomedical Laboratories, 5th Edition*. Available at: www.cdc.gov/biosafety/publications/bmbl5/BMBL.pdf.

[72] *See* You, E.H., op cit.

[73] NSABB, op cit.; Patterson, A., op cit.

[74] DHHS/NIH. (2009). *NIH Guidelines for Research Involving Recombinant DNA Molecules*. Available at: http://oba.od.nih.gov/rdna/nih_guidelines_oba.html.

[75] NSABB, op cit., p.13.

[76] Hart Research Associates, Synthetic Biology Project, Woodrow Wilson International Center for Scholars. (2010). *Awareness & Impressions of Synthetic Biology: A Report of Findings Based on a National Survey among Adults*. Available at: http://www.synbioproject.org/ process/assets/files/6451/_draft/hart_revised_.pdf.

[77] Framingham Heart Study. *Ethics Advisory Board*. Available at: http://www. framinghamheart study.org/about/ethics.html.

[78] NIH. *Genome-Wide Association Studies (GWAS)*. Available at: http://grants.nih.gov/grants/ gwas/.

[79] *See, e.g.*, Office of Human Research Protections. (2010) *Emergency Research Informed Consent Requirements*. Available at: http://www.hhs.gov/ohrp/humansubjects/guidance/ hsdc97-01.htm.

[80] NSABB, op cit..

[81] You, E.H., op cit.

[82] The charter of the President's Council on Bioethics instructed it "to provide a forum for a national discussion of bioethical issues." See Kass, L. (2005). Reflections on public bioethics: A view from the trenches. *Kennedy Institute of Ethics Journal* 15(3):221-250; The National Bioethics Advisory Commission frequently recommended ongoing review and assessment of the topics it studied, with its reports contributing to such a process. *See, e.g.*, National Bioethics Advisory Commission. (1999). *Ethical Issues in Stem Cell Research*. Available at: http://bioethics.georgetown.edu/nbac/pubs.html.

[83] For additional information, see Research Councils of the United Kingdom. *RCUK Public Engagement with Research Team*. Available at: http://www.rcuk.ac.uk/per/default.htm.

[84] Gutmann, A., and D. Thompson. (2002). Just deliberation about health care. In *Ethical Dimensions of Health Policy*. M. Danis, C. Clancy, and L. Churchill (eds). Oxford: Oxford University Press.

[85] *See, e.g.*, Sample, I. (2010). Craig Venter Creates Synthetic Life Form. *Guardian*. 20 May 2010. Available at: http://www.guardian.co.uk/science/2010/may/20/craig-venter-syntheticlife-form.

[86] Brenner, S., Senior Distinguished Fellow of the Crick-Jacobs Center, The Salk Institute. (2010). Reflections for the Future. Presentation to the Presidential Commission for the Study of Bioethical Issues, September 14, 2010. Available at: http://www.bioethics.gov/ transcripts/synthetic-biology/091410/reflections-for-the-future.html.

[87] Caplan, A., op cit.; Kaebnick,, G., (2009) op cit.; Wheeler, S.E., op cit.; Mattson, I., Director, Macdonald Center for the Study of Islam and Christian-Muslim Relations, Director, Islamic Chaplaincy Program, and Professor of Islamic Studies and Christian-Muslim Relations, Hartford Seminary; and President, Islamic Society of North America. (2010). Philosophical and Theological Perspectives. Presentation to the Presidential Commission for the Study of Bioethical Issues, September 14, 2010. Available at:

http://www.bioethics.gov/transcripts/synthetic-biology/091310/philosophical-and-theological-perspectives.html; Wolpe, P., Asa Griggs Candler Professor of Bioethics and Director, Center for Ethics, Emory University. Ethics of Synthetic Biology. Presentation to the Presidential Commission for the Study of Bioethical Issues, July 9, 2010. Available at: http://www.bioethics.gov/transcripts/syntheticbiology/070910/ethics-of-synthetic-biology.html; Van den Belt, H. (2009). Playing God in Frankenstein's footsteps: Synthetic biology and the meaning of life. *Nanoethics*. 3(3):257– 268; Dabrock, P. (2009). Playing God? Synthetic biology as a theological and ethical challenge. *Systems and Synthetic Biology* 3(1-4):47-54.

88 FactCheck.org. Available at: http://factcheck.org/.

89 The National Science Education Standards provides the following definition: "Scientific literacy means that a person can ask, find, or determine answers to questions derived from curiosity about everyday experiences. It means that a person has the ability to describe, explain, and predict natural phenomena. Scientific literacy entails being able to read with understanding articles about science in the popular press and to engage in social conversation about the validity of the conclusions. Scientific literacy implies that a person can identify scientific issues underlying national and local decisions and express positions that are scientifically and technologically informed. A literate citizen should be able to evaluate the quality of scientific information on the basis of its source and the methods used to generate it." National Committee on Science Education Standards and Assessment, National Research Council. (1996). *National Science Education Standards*. Washington, DC: National Academies Press, p.22.

90 Ibid.

91 For additional information about these programs, *see* Synthetic Biology Engineering Research Center. *Education*. Available at: http://www.synberc.org/content/articles/education; *see also* The Synthetic Biology Project of the Woodrow Wilson International Center for Scholars. Available at: http://www.synbioproject.org/.

92 For additional information, *see* BioEYES. *What is BioEYES?* Available at: http://www. bioeyes.org/.

93 National Bioethics Advisory Commission. (1999). *Research Involving Human Biological Materials: Ethical Issues and Policy Guidance*. Rockville, Maryland. Page 74. Available at: http://bioethics.georgetown.edu/nbac/pubs.html.

94 *See* Endy, D., Terman Fellow and Assistant Professor of Bioengineering, Stanford University. (2010). Overview and Context of the Science and Technology of Synthetic Biology. Presentation to the Presidential Commission for the Study of Bioethical Issues, July 8, 2010. Available at: www.bioethics.gov/transcripts/synthetic-biology/070810/overview-and-contextof-the-science-and-technology.html.

95 Rawls, J. (1971). *A Theory of Justice*. Cambridge, Mass.: Harvard University Press.

96 Harmon, A. (2010). "Indian Tribe Wins Right to Limit Research of its DNA." *New York Times*, April 21, 2010; and Couzin-Frankel, J. (2010). DNA returned to tribe, raising questions about consent. Science 328(5978):558.

97 45 C.F.R. § 46.111(a)(2).

98 Subcommittee on Environmental Justice, Environmental Health Policy Committee, DHHS. (1995). Special report: strategic elements for environmental justice. Environmental Health Perspectives 103(9):796-801.

99 Ibid.

100 Swiss Confederation, op cit.

101 Grove, N.E., et al. (2010). *The artemisinin enterprise: New sources of artemisinin to reduce economic barriers to ACT Access*. Available at: http://www.oneworldhealth.org/img/pdfdownloads/Artemisinin%20 Enterprise.pdf.

In: Synthetic Biology
Editors: Roger E. Hamilton and Laura M. Morales
ISBN: 978-1-62257-207-6

Chapter 2

ADDRESSING BIOSECURITY CONCERNS RELATED TO SYNTHETIC BIOLOGY*

National Science Advisory Board for Biosecurity

ABBREVIATIONS AND ACRONYMS

BSO	Biological Safety Officer
IBC	Institutional Biosafety Committee
NIH	National Institutes of Health
NIH Guidelines	*NIH Guidelines for Research Involving Recombinant DNA Molecules*
RAC	NIH Recombinant DNA Advisory Committee
USG	United States Government

EXECUTIVE SUMMARY

The National Science Advisory Board for Biosecurity (NSABB) was established by the U.S. Government (USG) to provide advice, guidance, and leadership regarding the oversight of dual use life sciences research – that is, research with a legitimate scientific purpose that yields information or technologies that may be misused to pose a threat to public health or other aspects of national security. In this report, the NSABB addresses the biosecurity and dual use research concerns that may arise from work being conducted in the nascent field of synthetic biology. The report considers synthetic biology in the context of the NSABB's proposed oversight framework for dual use research[1] as well as the biosafety guidelines that are described in the *NIH Guidelines for Research Involving Recombinant DNA Molecules*. Specifically, it describes the assessment of any biosecurity concerns presented by the ability to synthesize new genes, biochemical pathways, and biological components with specified or novel properties, and ultimately by the design of genetic systems, devices, and organisms

* This is an edited, reformatted and augmented version of Report of the National Science Advisory Board for Biosecurity (NSABB), dated April 2010.

with specified functions. As part of this assessment, the report examines whether all such biosecurity concerns would be adequately addressed by current and proposed oversight frameworks.

The term “synthetic biology” is used and defined in a variety of ways within the scientific community. Indeed, the dynamic nature of science all but ensures that a precise definition of “synthetic biology” will remain elusive and will evolve over time. The power of synthetic biology, as more broadly envisioned, reflects the possibility of synthetic biological systems that are programmable, self-referential, and modular. The ability to reformat the modules (e.g., the use of altered genetic codes or unnatural amino acids) leads to increased information content and vastly expanded possibilities for new function. Functionally, two disparate experimental approaches can be used to describe synthetic biology: “top down” and “bottom up.” The goal of both is to create novel biological structures with predictable properties and functions. The top down approach is related to classical recombinant DNA approaches and involves the re-engineering of existing organisms or genomes for a defined purpose. Increasingly powerful methods for DNA synthesis and assembly have significantly accelerated re-engineering capabilities, however. The bottom up approach entails the assembly of biological components in a variety of novel ways. This approach attempts to assemble systems (both living and nonliving) that perform desired functions in a predictable manner. Additionally, synthetic biology can be viewed even more broadly to include several emerging areas of study (e.g., nanotechnology, biocomputation), and we note that the biosecurity risks and dual use potential of new technologies and scientific information, regardless of their fields of origin, will remain important considerations for all areas of science.

Synthetic biology is a rapidly evolving field, and, given its potential benefits to public health and national and economic security, research in these disciplines should be encouraged and maintained. There are still several uncertainties surrounding the relatively nascent field, however. For example, one aspect of synthetic biology that differentiates it from other scientific disciplines is the degree to which it relies and depends on the ability to predict biological characteristics from nucleic acid or protein sequences or structures. But considerable scientific advancement still is needed for a sound understanding of how sequence, structure, and biological context contribute to biological properties, underscoring the difficulty in predicting how new biological materials will function.

Compounding this uncertainty is the pace at which synthetic biology is evolving, the sheer amount of new information the field is producing, and the growing diversity and number of practitioners. New approaches that enable genomes to be manipulated, rearranged and engineered on a large scale provide the ability to generate novel organisms whose properties are unknown. As with all scientific fields, it is often impossible to predict which novel discoveries will be made, but with synthetic biology this seems particularly challenging as technologies and new applications are being rapidly developed by especially varied practitioners.

This issue of diverse practitioners is particularly relevant with respect to synthetic biology because the field itself is somewhat of an amalgamation of several different disciplines. Synthetic biology has been described as “engineering biology”[2] and includes traditional life scientists as well as engineers, chemists, materials scientists, computer modelers and others. Moreover, the allure of synthetic biology has attracted the private sector, students at all levels (including high school), and amateur scientists who may lack formal institutional affiliations. Together, these practitioners include individuals of different ages and dissimilar social and educational backgrounds who may not have been sensitized to ethical, social, and legal norms and expectations of the traditional life sciences research communities.

Diversity among researchers is valuable for the scientific enterprise. But, since current biosafety and biosecurity paradigms address life sciences research conducted at research institutions, there well may be gaps in oversight resulting from the large numbers of synthetic biology practitioners who come from backgrounds that are not traditionally considered life sciences or who lack formal institutional affiliations. This highlights the importance, and the challenge, of raising awareness about dual use research issues and biosecurity concerns among individuals outside the life sciences, within the private sector and unaffiliated with research institutions.

Given these considerations for biosafety, biosecurity, and the dual use aspects of synthetic biology, the NSABB has developed the following recommendations:

1) Synthetic biology should be subject to institutional review and oversight since some aspects of this field pose biosecurity risks. Biosafety concerns can be adequately addressed by the application of current biosafety practices and procedures. If the proposed amendments to the *NIH Guidelines* are implemented, they will address research with synthetic nucleic acids in a more explicit fashion, including the expansion of guidance on risk assessment and risk management to address unique aspects of synthetic biology. To the extent that synthetic biology may present biosecurity or dual use research concerns, the NSABB has proposed an oversight paradigm that should adequately address such issues, but there is currently no federal policy in place for the review and conduct of dual use research of concern. The NSABB strongly urges the federal government to develop and implement such policy.
2) Oversight of dual use research should extend beyond the boundaries of life sciences and academia. While it is likely that current and proposed oversight paradigms will more than adequately address the biosafety and biosecurity concerns presented by synthetic biology, gaps in oversight remain, primarily due to the large numbers of synthetic biology practitioners who come from backgrounds that are not traditionally considered life sciences or who lack formal institutional affiliations. Moreover, synthetic biology is just one example of an area of science that may pose some dual use research concerns and whose practitioners span multiple scientific disciplines. In other instances, research that is highly relevant to the life sciences or that has implications for public health, agriculture, and the environment may be conducted outside the life sciences. Finally, dual use research of concern is as likely to be conducted in the private and voluntary sectors as it is in academia and government laboratories, so oversight should be uniform and comprehensive.
3) Outreach and education strategies should be developed that address dual use research issues and engage the research communities that are most likely to undertake work under the umbrella of synthetic biology. A critical first step in extending the oversight of dual use research of concern will be raising awareness of the dual use issue among synthetic biology's diverse practitioners, especially among those that have not been participants in recent discussions on this topic. The focus should be raising awareness about the potential biological, dual use, and other public health

implications of their work as well as the need for considering and addressing any biosecurity risks during the conduct of the research. Education efforts should be developed that target synthetic biology researchers who are a) not subject to federal biosafety and biosecurity requirements (e.g., private sector), b) not formally affiliated with universities or research institutions, and c) students (at all levels). This may present significant challenges, particularly in reaching individuals who lack formal affiliations, but this is important nonetheless.

4) The US Government should include advances in synthetic biology and understanding of virulence/pathogenicity in efforts to monitor new scientific findings and technologies, such as "tech-watch" or "science-watch" endeavors. It is appropriate for tech-watch or science-watch activities to identify emerging dual use technologies and new knowledge that could change the assessments of dual use risks and biosecurity concerns. As necessary, the USG should convene workshops to assess or re-assess our ability to create novel or unanticipated types of pathogens and to assess the biosecurity risks or dual use aspects of new technologies and whether they are adequately addressed by the extant biosecurity/dual use research oversight system.

INTRODUCTION

Purpose of This Document

The National Science Advisory Board for Biosecurity (NSABB) was established to advise the United States Government on strategies for the biosecurity oversight of dual use life sciences research. At its first meeting,[3] the NSABB was given a two-part charge concerning synthetic genomics and synthetic biology. The first part of the charge was to identify any potential biosecurity or dual use research concerns presented by the rapidly advancing ability to synthesize nucleic acids, especially the genomes of select agents.[4] Within that context, the NSABB was asked to assess the adequacy and applicability of the current regulatory and oversight framework for synthetic select agents. The findings and recommendations of the NSABB on this issue are set forth in a 2006 report, *Addressing Biosecurity Concerns Related to the Synthesis of Select Agents*.[5]

The second part of the charge, and the subject of this report, was to identify, assess and recommend strategies to address any biosecurity or dual use research concerns that may arise from work being conducted in the nascent field of synthetic biology. In accordance with its charter, the NSABB was to consider the potential that information and/or technology stemming from legitimate scientific research could be misused to threaten elements of national security. This report considers synthetic biology in the context of the NSABB's proposed oversight framework for dual use research[6] as well as the biosafety guidelines that are described in the *NIH Guidelines for Research Involving Recombinant DNA Molecules* (*NIH Guidelines*). Specifically, it describes the assessment of any biosecurity concerns presented by the ability to synthesize new genes, metabolic pathways, proteins, or chemicals and ultimately to design genetic systems and organisms with specified functions, and analyzes whether all such biosecurity concerns would be adequately addressed by current and proposed oversight frameworks.

What Is Synthetic Biology?

The term "synthetic biology" is used and defined in a variety of ways within the scientific community. For the purposes of this report, we use the term to encompass a variety of different goals and approaches, including:

- the design and construction of new biological parts and devices[7]—including computational devices, and other functional nucleic acid-based structures;
- the re-design of existing, natural biological systems for specific purposes;[8] as well as
- the synthesis of self replicating entities from scratch.[9]

The power of synthetic biology, as more broadly envisioned, reflects the possibility of synthetic biological systems that are programmable, self-referential,[10] and modular. The ability to reformat the modules (e.g., the use of altered genetic codes or unnatural amino acids) leads to increased information content and vastly expanded possibilities for new function. Two disparate experimental approaches are employed in synthetic biology. The goal of both is to create novel biological structures with predictable properties and functions. One is a "top down" approach that involves the re-engineering of existing organisms or genomes for a defined purpose. This might involve the metabolic engineering of microbes to produce useful products such as biofuels (or their precursors[11]) or synthetic molecules tailored for therapeutic purposes. This would also include synthesizing a life form that consists only of a minimal number of functional elements by removing redundant or non-essential genetic material, or that employs a modified genetic code. This so-called "chassis organism" would contain a genome that might be more easily engineered.12, 13, 14, 15 Another top down approach is to replace a portion of an existing genome with one or more standardized DNA modules that encode single, simple functions.[16] These approaches have made possible the creation of organisms with novel collections of properties. In fact, top down approaches for the re-engineering of microbes first appeared with the emergence of recombinant DNA techniques in the 1970s. Today, "classical" recombinant techniques and synthetic approaches are interwoven in the day-to-day practice of biological engineering. Yet the increasingly powerful methods for DNA synthesis and assembly have significantly accelerated re-engineering capabilities. Laborious, complex, site-directed mutagenesis and ligation procedures are now obviated by directed synthesis of the desired sequence, followed by relatively simple cloning or splicing steps. We use the term "synthetic biology" broadly and view practitioners as those who use classical recombinant molecular biology techniques as well as synthetic DNA constructs as we consider the impact and biosecurity implications of this discipline.

The other experimental approach in synthetic biology is to design and synthesize life forms, materials, or structures from the "bottom up" by taking non-living biological components and putting them together in a variety of novel ways. This approach assembles systems (both self-replicating and non-self-replicating) that perform desired functions in a predictable manner. Bottom up synthetic biology might involve creating artificial cells from nonliving materials with the goal of creating an entity that can regenerate, replicate, and evolve or that could be more easily engineered at the metabolic level. Bottom up synthetic biology may also include biofabrication – the synthesis of new materials or structures by assembling biological components such as nucleic acids, proteins or polysaccharides, perhaps in combination with non-biological components, into novel configurations. As envisioned, the synthetic or artificial life forms synthesized using this bottom up approach will not necessarily resemble or function like extant living cells. Likewise, when assembled, novel biomaterials may function in a different manner than would the material's individual components in their native biological context. The goal is to understand the fundamental nature of living organisms or biological materials and develop technology based on the same principles as those found in living systems.

With both approaches, the synthetic biologist seeks to understand the form and function of living organisms or their products (e.g., metabolites, enzymes, toxins) and utilize them in a predictable and controlled manner. Some describe this endeavor as being more akin to engineering than to biology because it involves taking parts of natural biological systems, characterizing and simplifying them, and using them as components of an unnatural, engineered, biological system.[17] Because of this, synthetic biology is sometimes referred to as engineering biology.[18]

One of the hallmarks of synthetic biology is that it is interdisciplinary in nature. In addition to biologists, practitioners also include engineers, chemists, and computer modelers. In a review article,[19] one of those practitioners, Drew Endy, describes the allure of synthetic biology for different types of scientists:

> ... for biologists, the ability to design and construct synthetic biological systems provides a direct and compelling method for testing our current understanding of natural biological systems; disagreements between expected and observed system behavior can serve to highlight the science that is worth doing. For chemists ... synthetic biology is an extension of synthetic chemistry; the ability to create novel molecules and molecular systems allows the development of useful diagnostic assays and drugs, expansion of genetically encoded systems, study of the origins of life, and so on. For "re-writers," the designs of natural biological systems may not be optimized for human intentions (for example, scientific understanding, health, and medicine); synthetic biology provides an opportunity to test the hypothesis that the genomes encoding natural biological systems can be "re-written," producing engineered surrogates that might usefully supplant some natural biological systems. Finally, for engineers, biology is a technology; building upon past work in genetic engineering, synthetic biology seeks to combine a broad expansion of biotechnology applications with ... an emphasis on the development of foundational technologies that make the design and construction of engineered biological systems easier.

In part, it is this interdisciplinary nature that makes synthetic biology such a promising field because individuals with related interests and diverse expertise can converge on a problem or challenge to produce rapid and profound results.

Finally, "synthetic biology" can be viewed even more broadly to include several emerging areas of study. For example, certain areas often cited as "nanotechnology" aim to engineer viral particles to serve as delivery devices for therapeutics.[20]

This type of research clearly incorporates engineering and biological aspects and may be considered synthetic biology by some. Electrical and computer engineers are beginning to develop DNA-based nanoelectronics and "biocomputation" applications21, [22] that further merge engineering and biotechnology and expand the boundaries of what is considered "synthetic biology."

The dynamic nature of science all but ensures that a precise definition of "synthetic biology" will remain elusive and evolve over time. However, the biosecurity risks and dual use potential of new technologies and scientific information, regardless of their fields of origin, will remain important considerations for all areas of science.

NSABB Approach

To examine the biosecurity concerns posed by synthetic biology and to determine whether they are adequately addressed by current oversight paradigms, the NSABB formed a Working Group on Synthetic Biology. The NSABB also hosted a scientific roundtable[23] to explore the state of the science, current capabilities, and future directions. (See Appendix B). Presenters were asked to describe their area of research and also to speak to following:

- how they define synthetic biology and what makes it unique among scientific approaches;
- the goals and experimental approaches in synthetic biology;
- the current capabilities and applications associated with synthetic biology;
- how these capabilities go beyond what is achievable using recombinant DNA or other related technologies;
- the major milestones to date in synthetic biology and the current challenges;
- future directions and goals;
- how close we are to predicting the detailed behavior of a cell based on its component parts;
- how close we are to designing biological systems and novel organisms with predictable functions; and
- whether their synthetic biology research routinely undergoes biosafety review.

The roundtable was co-hosted by the NIH Recombinant DNA Advisory Committee (RAC).[24] The topic of synthetic biology was of interest to this group as well because the RAC was tasked with assessing the applicability of the *NIH Guidelines* to synthetic genomics and synthetic biology. This tasking was the result of the US government acting on one of the recommendations from the NSABB regarding the need to ensure that biosafety guidance and requirements adequately address work with synthetic nucleic acids and are understood by all those working in areas subject to the biosafety guidance and requirements.[25]

Towards this end, one session of the roundtable was focused on current understanding of the relationship between biological properties and sequence and structure, the current ability to predict biological properties, and the tools that are available for predicting function.

The Working Group also received a briefing in September 2009 by the Executive Director of the RAC, Dr. Jacqueline Corrigan-Curay, on proposed updates to the *NIH Guidelines* that address synthetic nucleotides and organisms.

In addition, the Working Group received a briefing by Special Agent Edward You regarding the recent outreach efforts of the Federal Bureau of Investigation (FBI) aimed at the synthetic biology community and a 2009 FBI-sponsored conference titled "Building Bridges Around Building Genomes."

Throughout the course of the Working Group's deliberations on this topic, NSABB members provided updates to the group after informally attending meetings (including the FBI conference above) and symposia. NSABB members also informally consulted with subject-matter experts during the process to help inform the group's findings and recommendations.

Because the use of synthetic oligonucleotides is often combined with the use of naturally occurring genetic material and involves recombinant DNA methods and approaches, this report is directed at a broad range of science and scientists and not just those who work exclusively with synthetic DNA.

FINDINGS

The Promise of Synthetic Biology: Early Days in a Rapidly Evolving Field

During the course of the NSABB's deliberations, it became evident that synthetic biology is both a relatively nascent and rapidly developing field. Although the term "synthetic biology" was coined in 1912 by Stéphane Leduc in reference to the synthesis of artificial life,[26] the field of synthetic biology as a discipline within the life sciences remains in its relative infancy.

However, in recent decades, synthetic biology has benefited from advances in related disciplines, allowing this burgeoning field to develop into one that holds great promise to improve human, animal, plant, and environmental health Advances in molecular biology have allowed genetic material to be readily manipulated, enabling the synthetic biologist to splice virtually any piece of DNA into another. High-throughput DNA sequencing has provided the synthetic biologist with access to the complete genome sequences of more than 980 microorganisms[27] and advances in DNA synthesis technology allows the rapid chemical synthesis and assembly of suites of genes (e.g., the genes encoding a metabolic pathway) *de novo*. A deeper understanding of the specific types and functions of RNA molecules has allowed the synthetic biologist to construct, and program, biological devices.[28] Moreover, advances in other fields such as materials science and protein engineering are allowing the fabrication of new biomaterials.

While many of the ambitious goals of synthetic biology have yet to be achieved (e.g., *de novo* synthesis of a wholly new organism or the ability to reliably assemble biological components into complex systems with predictable outcomes), in recent years the science has matured to a point where some of this field's promise is being realized. Synthetic approaches have resulted in the more efficient production of therapeutics that are less expensive and more environmentally benign,[29, 30] as well as the development of new materials for tissue and organ growth and drug delivery.[31, 32, 33] More efficient means of producing energy are being developed that utilize metabolically engineered microbes that can convert plentiful, renewable resources into hydrogen, ethanol, diesel, or biofuel precursors.[34, 35] In an approach referred to as "synthetic metagenomics," automated chemical synthesis and computational optimization of genes from environmental metagenomic libraries have been used to create symbiotic consortia of engineered organisms that produce significant amounts of precursor molecules for important industrial chemicals and fuels.[36] Plants and microbes also are being engineered for biological remediation applications that offer low-cost, environmentally friendly solutions to pollution and contamination.[37, 38] In addition, synthetic approaches have been employed toward the detection and neutralization of, and defense against chemical and biological threats. Given the great number of potential benefits offered by synthetic biology, encouraging and maintaining a strong research program in this discipline is crucial to the nation's biological, physical and economic security.

Significant Uncertainties

Presentations and discussion at the NSABB-RAC roundtable were particularly helpful for understanding the multiplicity of goals and approaches in synthetic biology, the state of the science of systems biology, and the ability to predict key biological properties such as virulence from nucleic acid sequence. One aspect of synthetic biology that differentiates it from other disciplines is the degree to which it relies and depends on the ability to predict biological characteristics from nucleic acid or protein sequences or structures. This is especially true for the bottom up approach where new organisms and/or structures are assembled using more simple starting materials. However, with all aspects of synthetic biology, there are varying degrees of uncertainty regarding the predictability of biological properties of partially or completely synthetic agents or materials. This is due in large part to the fact that the biological context or environment of a gene or gene product is absolutely key to its function. Biological environment(s) may in fact be as important a determinant of the function of a biological component of that environment, as is the sequence or structure of the component.

As a follow-up to a recommendation made by the NSABB in its 2006 report on synthetic genomics,[39] the US Government commissioned the National Research Council to examine the uncertainties surrounding the ability to predict biological function from sequence in a project titled *Scientific Milestones for the Development of a Gene-Sequence-Based Classification System for Oversight of Select Agents*. An aim of this study is "to identify the scientific advances that would be necessary to permit serious consideration of developing and implementing an oversight system for select agents that is based on predicted features and properties encoded by nucleic acids rather than a relatively static list of specific agents and taxonomic definitions."[40] While this project focuses on the scientific advancements that will be required to develop a predictive oversight framework for select agents, it underscores the desirability of and current limitations to accurately predicting biological characteristics from DNA sequences. Clearly, the Council's findings will be highly relevant to the field of synthetic biology.

However, considerable scientific advancement is still needed for a sound understanding of how sequence, structure, and biological context contribute to biological properties. Until those scientific milestones are realized, it will continue to be difficult to predict with any reasonable certainty the biological risk of a synthetic entity, especially one that bears little resemblance to natural organisms. This inability to predict function from sequence calls for greater awareness of potential biosafety and biosecurity risks, screening and characterizing the properties of DNA constructs and engineered proteins and organisms, and greater sharing of practices among the scientific community. Of note, a recurring theme in the NSABB's discussions with experts was that, although synthetic biology poses potential risks, there remain significant limitations to our current ability to custom design and produce novel organisms, especially pathogens, either by *de novo* synthesis or by engineering extant organisms in predictable ways.

Compounding this uncertainty is the pace at which synthetic biology is evolving and the sheer amount of new information the field is producing. Massively parallel DNA sequencing and synthesis approaches are increasing the speed and decreasing the cost of synthesizing virtually any custom DNA sequences *de novo*.[41] New approaches have been developed to engineer entire microbial genomes as well, using synthetic oligonucleotides that integrate into multiple chromosomal locations in a rapid, continuous manner, generating combinatorial genomic diversity and accelerated evolution.[42] In a demonstration of this approach, investigators produced up to 15 billion genetic variants, and of this mutagenized population, cells were selected and characterized with enhanced efficiency in expressing the products of a particular metabolic pathway. Other approaches have been developed to drive chromosomal evolution and engineer biosynthetic pathways in microbes.[43] The ability to mutagenize microbial populations is not new and has been employed routinely by researchers. However, new approaches that enable genomes to be manipulated and rearranged on a large scale provide the ability to generate novel organisms whose properties are unknown. The possibility that, during the research process, certain selective pressures might lead to the identification of mutants with enhanced virulence presents potential biosafety, biosecurity and/or dual use concerns.

Finally, the uncertainties surrounding synthetic biology are amplified further by the myriad of potential applications being developed by a growing number of practitioners. Synthetic biology is a rapidly developing field, and, as noted above, it is conducted not only by biologists but also by engineers, chemists, material scientists and others, often through interdisciplinary collaborations. It is impossible to predict the information, technologies, and new applications that will be developed by or applied to this relatively new field that is attracting especially diverse practitioners.

In light of this, it is necessary to conduct synthetic biology research under appropriate biosafety and biosecurity oversight. This may require a risk assessment, which may be challenging depending on the nature of the work. For instance, the degree of novelty of the synthetic agent and how closely its biological properties might track to a parental agent(s) may need to be considered. Of particular concern, in terms of a safety and security risk assessment, are experimental techniques, such as directed molecular evolution ("DNA shuffling") and shotgun synthesis that generate very large numbers of recombinants and sequence variants.[44] While it is likely that only a very small fraction of resulting new organisms might be viable, it would be very difficult, if not impossible, to predict biological properties, including virulence. In addition, as with all biological materials, any material derived from synthetic means should be disposed of in the proper manner.

Current Oversight Paradigms

The NSABB considered how potential biosafety and biosecurity concerns posed by synthetic biology would fit within current or proposed oversight paradigms and whether any gaps might exist. Currently, basic or clinical research involving recombinant DNA being conducted at NIHfunded institutions is subject to the *NIH Guidelines*. The *NIH Guidelines* outline principles for safe research practices with respect to recombinant DNA molecules.

They detail sound practices, including procedures for the containment of various forms of recombinant DNA research, including research involving genetically modified microorganisms, plants and animals, as well as human gene-transfer experiments. Institutions must also establish an Institutional Biosafety Committee (IBC) to review research involving recombinant DNA. Under certain circumstances the *NIH Guidelines* require that institutions appoint a Biological Safety Officer (BSO) to oversee the management of biosafety risks. In addition, the Recombinant DNA Advisory Committee (RAC) was established to advise the NIH Director on the content and implementation of the *NIH Guidelines* and to provide public, in-depth review of the scientific, safety, and ethical dimensions of research and clinical trials that involve the transfer of recombinant DNA to humans.[45]

In 2009, the RAC proposed expanding the scope of the *NIH Guidelines* to include nucleic acid molecules made solely through synthetic means.[46] In general, the RAC has found that, in most cases, research with synthetic nucleic acids presents biosafety risks that are comparable to recombinant DNA research and that certain work with synthetic nucleic acids in a non-replicating form may not require oversight under the *NIH Guidelines* (although other biosafety standards would apply). It also found that the current risk assessment framework described in the *NIH Guidelines* can be used to evaluate synthetically produced nucleic acids with attention to the unique aspects of this technology. The RAC also has noted how rapidly the field is growing and concluded that safety issues surrounding synthetic nucleic acids will likely need to be revisited in the near future as proposed amendments to the *NIH Guidelines* can only address what is presently known about the science.

Previously, the NSABB has proposed an oversight framework for addressing dual use research concerns for life sciences research.[47] The framework is based on a proposed criterion for identifying research that constitutes "dual use research of concern", and describes principles that should underpin the oversight of dual use life sciences research, key features of such an oversight system, and strategies for assessing and managing the risks posed by dual use research of concern. The NSABB's proposed framework relies primarily on the local oversight of dual use research. Briefly, the NSABB envisions a system wherein principal investigators make the initial evaluation (and periodic re-evaluation) of their work for its dual use potential. Research identified as being potentially of dual use concern would receive additional institutional review and risk-management plans would be developed as appropriate (and would include plans for the responsible communication of research findings). Researchers are the most critical element in the NSABB's oversight framework because they are in the best position to anticipate the types of information or technologies that might be generated by their work. As such, the NSABB emphasized that awareness of the dual use issue must be the foundation of such a system, and accordingly has developed outreach and educations strategies that target relevant communities.[48] There is currently no federal policy for the oversight of dual use research; the NSABB's oversight framework is being considered by the US Government, however, and the recommendations contained in this report assume that a federal policy is forthcoming.

Biosafety and Biosecurity Concerns

The *NIH Guidelines* and the NSABB's proposed framework for dual use research oversight focus on individuals participating in life sciences research and operating within a university or institutional setting. Not all biologists operate within these settings, however, and this is especially true of practitioners of synthetic biology. In addition, despite the term synthetic "biology," not all practitioners consider that their work is biological in nature; rather, they may view their work as a type of engineering, an extension of synthetic chemistry, or materials science.[49] As such, they may not be considering the biological and public health implications of their work. In addition, many practitioners have backgrounds that are not rooted in the life sciences. Consequently, their training may not have included or emphasized principles and practices of biological risk assessment and biocontainment. Thus, raising awareness within the disparate scientific communities that engage in synthetic biology about the possible biosafety risks and the need for the responsible conduct of research will be critical.

We note that biosafety and biosecurity are two distinct but related concepts.[50] Biosafety typically refers to the policies, practices, equipment, facilities, and medical treatments designed to protect workers and the environment from the accidental exposure to hazardous laboratory agents and materials. Biosecurity refers to the protection, control of, and accountability for high-consequence biological agents and toxins, and critical relevant biological materials and information, to prevent unauthorized possession, loss, theft, misuse, diversion, or intentional release. The concept of 'dual use research of concern' is an aspect of biosecurity because it refers to the potential for misuse of scientific information to threaten public health, animal or plant populations, or other aspects of national security. While the focus of the NSABB is biosecurity, biosafety and biosecurity often converge because both require the assessment and management of research-associated risks.

The biosecurity or dual use research concerns potentially associated with synthetic biology might be considered according to experimental goal and approach. The bottom up synthetic biology research that aims to synthesize artificial life forms from scratch is still in its infancy and likely does not pose any significant biosecurity or dual use research concerns at the present time that would not be addressed, in theory, by the proposed oversight system of dual use research. Biosafety concerns regarding recombinant techniques typically utilized in this type of research would be adequately covered by the *NIH Guidelines*. However, since bottom up approaches toward developing synthetic organisms may be conducted by individuals without a background in the life sciences, or outside the formal settings subject to biosafety and dual use research oversight, the proposed outreach and education programs, from a practical perspective, may not reach a significant number of practitioners.

Well-developed and ongoing bottom up approaches also include the design, construction, and use of new biological parts, devices, and systems. Biological devices, switches, and engineered gene circuits that can respond to stimuli and behave in a predicted manner have been developed using simple biological materials.51, [52] Moreover, efforts to standardize biological parts are beginning to enable interchangeability. For example, standardized DNA sequences with defined structures and functions are being developed such that they share a common interface and can be readily spliced together with the aim of designing and/or programming new, living biological systems. A "parts-type" sequence might encode basic biological functions such as encoding a certain protein or providing a promoter to let RNA polymerase bind and initiate transcription of downstream sequences. "Device-type" sequences are collections of parts that implement some human-defined function such as producing a fluorescent protein whenever the environment contains a certain chemical. "Systems-type" sequences perform high-level tasks such as oscillating between two colors at a predefined frequency or serving as a toggle switch. A registry of several hundred of these standardized parts is maintained in the public domain to facilitate the open sharing of these interchangeable pieces of DNA.[53] Whereas the standardization of biological parts itself does not present biosecurity concerns, the particular applications of them could. For example, it might be possible to design an organism that contains a biological switch for the production of a toxin. Although this could be considered dual use research of concern, it also should be adequately addressed by researchers and institutions while they assess their recombinant research for safety risks and for dual use potential.

Top down approaches, namely the re-design of existing, natural biological systems for specific purposes, could in some circumstances pose dual use or biosecurity risks. Obvious examples are the deliberate creation of novel pathogens, enhancement of the pathogenicity of a naturally occurring pathogen, and the re-design of a non-pathogen into a pathogen using synthetic biology technologies. Again, however, we anticipate that the safety concerns presented by these types of experiments would be adequately addressed under the *NIH Guidelines* and the dual use risks would be addressed by the oversight paradigm previously proposed by the NSABB (aside from the issue of adequate outreach to these practitioners). Indeed, the NSABB anticipated these types of experiments, and others of potential dual use concern, in the broader context of life sciences research when developing the proposed oversight framework for dual use research.[54]

In all of these cases, as the science and technology progresses, it will be necessary to reconsider both the biosecurity and the biosafety issues. It also should be emphasized that the current systems of oversight are aimed at life sciences practitioners within universities or institutional settings; those individuals conducting synthetic biology research without a life sciences background or formal affiliations may present a gap in the current oversight systems.

Implications of Diverse Practitioners

Another potential concern—that perhaps falls somewhere between biosafety and biosecurity— involves the diversity of the individuals conducting synthetic biology research. As synthetic biology techniques become easier and less expensive and the applications become more widely relevant, the range of practitioners expands to include scientists from a variety of disciplines; students at all levels, including high school; and amateur scientists and hobbyists who may lack any formal affiliations with universities or research institutions. The diversity of practitioners will also include individuals of different ages and varied social and educational backgrounds who may not have been sensitized to the ethical, social and legal norms of the traditional life sciences research communities. However, these communities have a longstanding engagement with such concerns. For example, in 1975, experts convened to discuss the risks and benefits of the newly emerging field of recombinant DNA research at the Asilomar Conference on Recombinant DNA Molecules. This landmark conference illustrates these communities' commitment to the responsible conduct of research. The importance of its legacy should be impressed on the new generation of researchers, including those not affiliated with the life sciences.

Certainly, students at all levels who are utilizing synthetic biology technology need to be educated about the importance of working responsibly. The importance of educating students about safety has been recognized within some components of the synthetic biology community, and, as one example, this year, participants in the International Genetically Engineered Machine (iGEM) competition were required to address biosafety issues as part of their projects.[55] At the same time, to hold educational activities to the same biosafety and biosecurity oversight paradigm as academic and industry research probably is unrealistic and may have the unintended effect of stifling some excellent opportunities for getting young people excited about science.

Thus, education efforts and oversight requirements should be tailored to the audience. For the practitioners of synthetic biology who are from non-biological disciplines, outreach and education efforts will need to be targeted and designed differently from those directed at scientists who already are familiar with biosafety and biosecurity principles and practices.

Effectively engaging amateur scientists and hobbyists—in terms of both education and oversight—is even more challenging but nonetheless needs to be addressed, even as these "citizen scientists" are forming communities that are engaged with biosafety issues and responsible oversight and developing codes of ethics.[56] In fact, for those with little history of discussion of these issues, biosafety may become confused with or a surrogate for biosecurity, and prevent proper consideration of the latter. Sensitivity about the appropriate physical management of biological materials is not the same thing as sensitivity about the information generated from the experimental use of the same materials.

Although the variety of synthetic biology practitioners presents a challenge to current oversight systems, this diversity is good for the scientific enterprise. Such diversity enables the convergence of expertise and perspectives that leads to fruitful collaborations and exciting new findings and technologies.

Synthetic biology, as a discipline, is poised to grow, and the increased participation of individuals with many different interests is inevitable. This highlights the importance, and the challenge, of raising awareness about dual use research and biosecurity issues among individuals outside the life sciences, within the private sector and unaffiliated with research institutions.

The Importance of Engagement About Synthetic Biology

The goals, potential benefits and risks, and current limitations of synthetic biology are not uniformly understood within the scientific community or the general public. More effective and extensive dialogue within and across the disparate scientific communities engaged in synthetic biology is necessary in this regard. Certainly, more effective outreach to the general public is needed to foster a more realistic understanding of synthetic biology. Public engagement would shed light on the goals, as well as the current limitations of synthetic biology, and would not only generate excitement about this research but help to dispel some misconceptions and concerns that designer pathogens are readily available.

RECOMMENDATIONS

Synthetic biology is an emerging field with tremendous potential to impact in a positive manner the overall health and environment of not only the US, but the planet. As such, it is important to encourage the further development of this scientific endeavor and not place undue restrictions on the ability to move the research forward. At the same time, there are some potential biosecurity risks associated with synthetic biology, and in particular, the concern that synthetic biology offers an opportunity for the synthesis or re-design of harmful pathogens that could be used to threaten public, plant or animal health. Another, potentially larger, risk is the development of information or technology through the normal scientific process that could be misused to threaten aspects of national security. With these considerations in mind, we recommend the following:

1) Synthetic biology should be subject to institutional review and oversight since some aspects of this field pose biosecurity risks. Biosafety concerns can be adequately addressed by the application of current biosafety practices and procedures. If the proposed amendments to the *NIH Guidelines* are implemented, they will address research with synthetic nucleic acids in a more explicit fashion, including the expansion of guidance on risk assessment and risk management to address unique aspects of synthetic biology. To the extent that synthetic biology may present biosecurity or dual use research concerns, the NSABB has proposed an oversight paradigm that should adequately address such issues, but there is currently no federal policy in place for the review and conduct of dual use research of concern. The NSABB strongly urges the federal government to develop and implement such policy.
2) Oversight of dual use research should extend beyond the boundaries of life sciences and academia. While it is likely that current and proposed oversight paradigms will more than adequately address the biosafety and biosecurity concerns presented by synthetic biology, gaps in oversight remain, primarily due to the large numbers of synthetic biology practitioners who come from backgrounds that are not traditionally considered life sciences or who lack formal institutional affiliations. Moreover, synthetic biology is just one example of an area of science that may pose some dual use research concerns and whose practitioners span multiple scientific disciplines. In other instances, research that is highly relevant to the life sciences or that has implications for public health, agriculture and the environment may be conducted outside the life sciences. Finally, dual use research of concern is as likely to be conducted in the private and voluntary sectors as it is in academia and government laboratories, so oversight should be uniform and comprehensive.
3) Outreach and education strategies should be developed that address dual use research issues and engage the research communities that are most likely to undertake work under the umbrella of synthetic biology. A critical first step in extending the oversight of dual use research of concern will be raising awareness of the dual use issue among synthetic biology's diverse practitioners, especially among those that have not been participants in recent discussions on this topic. The focus should be raising awareness about the potential biological, dual use, and other public health implications of their work as well as the need for considering and addressing any

biosecurity risks during the conduct of the research. Education efforts should be developed that target synthetic biology researchers who are a) not subject to federal biosafety and biosecurity requirements (e.g., private sector), b) not formally affiliated with universities or research institutions, and c) students (at all levels). This may present significant challenges, particularly in reaching individuals who lack formal affiliations, but this is important nonetheless.

4) The US Government should include advances in synthetic biology and understanding of virulence/pathogenicity in efforts to monitor new scientific findings and technologies, such as "tech-watch" or "science-watch" endeavors. It is appropriate for tech-watch or science-watch activities to identify emerging dual use technologies and new knowledge that could change assessments of dual use risks and biosecurity concerns. As necessary, the USG should convene workshops to assess or re-assess our ability to create novel or unanticipated types of pathogens and to assess the biosecurity risks or dual use aspects of new technologies and whether they are adequately addressed by the extant biosecurity/dual use research oversight system.

APPENDIX A: NATIONAL SCIENCE ADVISORY BOARD FOR BIOSECURITY ROSTER

Members of the NSABB Working Group on
Synthetic Biology are denoted with an asterisk (*)

Chair

Dennis L. Kasper, M.D.
William Ellery Channing Professor of Medicine
and Professor of Microbiology and Molecular
Genetics
Harvard Medical School;
Director, Channing Laboratory
Dept. of Medicine
Brigham and Women's Hospital
Boston, MA

Other Voting Members

Arturo Casadevall, M.D., Ph.D.
Professor and Chairman
Dept. of Microbiology and Immunology
Albert Einstein School of Medicine Bronx, NY

Murray L. Cohen, Ph.D., M.P.H., C.I.H.
President and Chairman
Frontline Healthcare Workers®
Safety Foundation, Ltd. Atlanta, GA
Susan A. Ehrlich, J.D., LL.M.*
Judge (Retired)
Arizona Court of Appeals
Phoenix, AZ

Lynn W. Enquist, Ph.D.
Professor and Chair
Dept. of Molecular Biology
Princeton University;
Editor and Chief,
Journal of Virology Princeton, NJ

Barry J. Erlick, Ph.D.
President
BJE Associates, Inc.;
Affiliate Professor
Auburn University;
'Adjunct Professor
Kansas State University
Alexandria, VA

David R. Franz, D.V.M., Ph.D.
Vice President and Chief Biological Scientist
Midwest Research Institute;
Director, National Agricultural Biosecurity Center
Kansas State University
Frederick, MD

Claire M. Fraser-Liggett, Ph.D.*
Director, Institute of Genome Sciences
University of Maryland School of Medicine
Baltimore, MD

General John A. Gordon
General, USAF (Retired)
Alexandria, VA

Michael J. Imperiale, Ph.D.*
Professor
Dept. of Microbiology and Immunology
University of Michigan Medical School
Ann Arbor, MI

Paul S. Keim, Ph.D.
Division Director, Pathogen Genomics
Translational Genomics Research Institute;
Cowden Endowed Chair in Microbiology
Northern Arizona University
Flagstaff, AZ

Stanley M. Lemon, M.D.
Director
Institute for Human Infections and Immunity
University of Texas Medical Branch
at Galveston
Galveston, TX

Stuart B. Levy, M.D.
Director
Center for Adaptation Genetics and Drug
Resistance;
Professor of Medicine and Molecular Biology
and Microbiology
Tufts University School of Medicine
Boston, MA

John R. Lumpkin, M.D., M.P.H.
Senior VP and Director of the Health Care Group
Robert Wood Johnson Foundation Princeton, NJ

Adel A.F. Mahmoud, M.D., Ph.D.
Professor, Dept. of Molecular Biology a
nd Woodrow Wilson School
Princeton University; Princeton, NJ

Mark E. Nance, J.D.
General Counsel
Medical Diagnostics
GE Healthcare
Princeton, NJ

Michael T. Osterholm, Ph.D., M.P.H.
Director, Center for Infectious
Disease Research and Policy;
Associate Director, Dept. of Homeland Security
National Center for Food Protection and Disease;
Professor, School of Public Health
University of Minnesota
Minneapolis, MN

David A. Relman, M.D.*
Chair, NSABB Working Group on Synthetic Biology
Thomas C. and Joan M. Merigan Professor
Departments of Medicine, and of Microbiology & Immunology
Stanford University School of Medicine Stanford, CA

James A. Roth, D.V.M., Ph.D., D.A.C.V.M.
Director, Center for Food Security and Public Health
Executive Director, Institute for International
Cooperation in Animal Biologics
College of Veterinary Medicine
Iowa State University
Ames, IA

Harvey Rubin, M.D., Ph.D.*
Professor of Medicine
University of Pennsylvania School
of Medicine Philadelphia, PA

Thomas E. Shenk, Ph.D.*
James A. Elkins, Jr. Professor in the Life Sciences
Dept. of Microbiology
Princeton University
Princeton, NJ
Andrew A. Sorensen, Ph.D.
Distinguished Professor of the University and
Distinguished President Emeritus
University of South Carolina
Columbia, SC

Admiral (Ret.) William O. Studeman
Consultant; VP Northrup Grumman Corporation
(Retired) Great Falls, VA

Anne K. Vidaver, Ph.D.
Professor and Head
Dept. of Plant Pathology
University of Nebraska-Lincoln
Lincoln, NE

Non-voting *Ex Officio* Members

Jason Boehm, Ph.D.
Office of the Director National Institute of
Standards and Technology Department of Commerce

Brenda A. Cuccherini, Ph.D., M.P.H.
Special Assistant to the Chief R&D Officer
Office of Research and Development
Veterans Health Administration
Department of Veterans Affairs

Diane C. DiEuliis, Ph.D.*
Assistant Director, Life Sciences
Office of Science and Technology Policy
Executive Office of the President

Anthony S. Fauci, M.D.
Director
National Institute of Allergy and Infectious
Disease
National Institutes of Health
Department of Health and Human Services

Elizabeth George, Ph.D.*
Division Head
Chemical and Biological Division Science and
Technology Directorate Department of Homeland Security

Sue D. Haseltine, Ph.D.*
Associate Director for Biology
U.S. Geological Survey
Department of the Interior

Tom Hopkins, Ph.D.
Assistant to the Secretary of Defense for Nuclear
and Chemical and Biological Programs (Acting)
Department of Defense

Peter R. Jutro, Ph.D.*
Deputy Director
National Homeland Security Research
Center Environmental Protection Agency

Boris D. Lushniak, M.D., M.P.H.
Chief Medical Officer
Office of the Commissioner
Office of Counter-terrorism Policy
Food and Drug Administration
Department of Health
and Human Services

Mary Mazanec, M.D., J.D.
Director, Office of Medicine, Science and Public Health
Office of the Assistant Secretary for Preparedness and Response
Department of Health and Human Services

Robert P. Mikulak, Ph.D.
Bureau of International Security and Nonproliferation
Office of Chemical and Biological Weapons
Threat Reduction
U.S. Department of State

Janet K. A. Nicholson, Ph.D.
Associate Director for Laboratory Science
National Center for Infectious Diseases
Centers for Disease Control and
Prevention Department of Health
and Human Services

Caird E. Rexroad, Jr., Ph.D.*
Associate Administrator
Agricultural Research Service
U.S. Department of Agriculture

Amanda Dion-Schultz, Ph.D.*
Office of the Chief Scientist

David G. Thomassen, Ph.D.*
Chief Scientist
Office of Biological & Environmental Research
Office of Science
Department of Energy

Joanne Tornow, Ph.D.*
Division Director (Acting)
Molecular and Cellular Biosciences
Directorate for Biological Sciences
National Science Foundation

Edward You*
Supervisory Special Agent
FBI Weapons of Mass
Destruction Directorate
Countermeasures Unit
Bioterrorism Program
Federal Bureau of Investigation
Department of Justice

Additional Federal Representatives

Kenneth Cole, Ph.D.*
CAPT, MSC, USN
Deputy and Medical Director
Office of the Special Assistant to the Secretary
of Defense for Chemical and
Biological Defense
and Chemical Demilitarization Programs
OSA (CBD&CDP)
Department of Defense

Daniel Drell, Ph.D.*
Program Manager
Biological Systems Sciences Division
SC-23.2/Germantown Bldg
Office of Biological and
Environmental Research
Office of Science
Department of Energy

Maria Giovanni, Ph.D.*
Assistant Director for Microbial Genomics
& Advanced Technology
Division of Microbiology and
Infectious Diseases
National Institute of Allergy and
Infectious Diseases
National Institutes of Health

Wendy Hall, Ph.D.*
Director
Bioterrorism, Science and Technology
Department of Homeland Security

David Holmes, Ph.D.*
Division of Select Agents and Toxins
Coordinating Office for Terrorism
Preparedness and Emergency Response
Jessica Petrillo, Ph.D.*
Microbiologist
Office of Chemical and Biological Weapons
Threat Reduction
Bureau of International Security and
Nonproliferation
Department of State

Jessica Tucker, Ph.D.*
AAAS Science and Technology Policy Fellow
Department of Health and Human
Services Office of the Assistant
Secretary for Preparedness and Response
Office of Medicine, Science,
and Public Health

Tru Twedt, D.V.M., C.B.S.P.*
Associate Director for Science
Division of Select Agents and
Toxins Centers for Disease
Control and Prevention

Robbin Weyant, Ph.D.*
Director, Division of Select Agents and Toxins,
Coordinating Office for Terrorism
Preparedness and Emergency Response
Centers for Disease Control and Prevention

Executive Director

Amy P. Patterson, M.D.
Executive Director, NSABB
Acting Director, Office of Science Policy
Office of the Director
National Institutes of Health

APPENDIX B: AGENDA
ROUNDTABLE ON SYNTHETIC BIOLOGY
OCTOBER 11, 2007

NIH Campus
9000 Rockville Pike, Bethesda, MD
Building 31, C Wing, 6th Floor, Conference Room 10

8:00 a.m. **Welcome and Opening Remarks**
Paul Keim, PhD NSABB Member
Northern Arizona University

Howard Federoff, M.D., Ph.D. Chair, NIH RAC
Georgetown University Medical Center

8:10 a.m. **Overview of Roundtable**

David Relman, M.D.
Chair, NSABB Working Group on
Synthetic Genomics Stanford University

8:20 a.m. **Session 1: State of the Science of Synthetic Biology**

This session will provide a broad understanding of the field of synthetic biology—the various research approaches, current capabilities, short and long-term goals. Questions to be addressed during the presentations and/or discussion period are listed below.

Moderators:

Harvey Rubin, M.D., Ph.D.
NSABB Member
Director, Institute for Strategic Threat Analysis and Response Professor of Medicine, Microbiology, and Computer Science University of Pennsylvania

Stephen Dewhurst, Ph.D. RAC Member
Professor of Microbiology & Immunology and of Oncology
University of Rochester

Questions for Consideration

How would you define the field of work to which various parties refer under the rubric of "synthetic biology"? How is this field of scientific activity related to, and distinguished from other fields? What are the goals of synthetic biology?

What are the principal lines of research in this area?

What are the capabilities and applications associated with the field of synthetic biology at present? What capabilities do the approaches used in synthetic biology provide that are beyond those that can be achieved using recombinant DNA or other related technologies?

What are some of the more important accomplishments to date by those that might be labeled as synthetic biologists? What are the current major hurdles or challenges in synthetic biology?

Where do you see the field of synthetic biology headed in the near and long terms?

How close are we to predicting the detailed behavior of a cell based upon its component parts? How close are we to designing biological systems and novel organisms with predictable functions?

Does your research within the realm of synthetic biology routinely undergo any biosafety review? If so, please describe.

8:25 a.m. **A Bird's Eye View of Synthetic Biology**

Roger Brent, Ph.D. Director and President
The Molecular Sciences Institute; Berkeley, CA

8:50 a.m. **Synthetic Biology As A Twenty Year Old Field**

Steven A. Benner, Ph.D.
Distinguished Fellow
Foundation for Applied Molecular Evolution Gainesville, FL

9:15 a.m. **Synthetic Biology: From Bacteria to Stem Cells**

Ron Weiss, Ph.D.
Professor of Electrical Engineering Princeton University

9:40 a.m. **Synthetic Organisms**

Steen Rasmussen, Ph.D.
Team Leader for Self-Organizing Systems
Los Alamos National Laboratory

10:05 a.m. ***Break***
10:20 a.m. **Discussion**

11:40 a.m. Break for Working Lunch

12:00 noon **Session 2: Predicting Function**

This session will address our current understanding of the relationship between biological properties and sequence and structure, our ability to predict biological function/properties, and the tools that are available or under development for predicting function. Questions to be addressed during the presentations and/or discussion period are listed below.

Moderators:

Claire Fraser-Liggett, Ph.D.
NSABB Member
Director, Institute of Genome Sciences
University of Maryland School of Medicine, Baltimore

Nikunj Somia, Ph.D.
RAC Member
Assistant Professor
University of Minnesota, Twin Cities

Questions for Consideration

To what degree are we able to explain virulence of the more common bacterial and viral pathogens, based on their genome sequences? How accurately can virulence or other pathogenic properties be predicted on the basis of sequence alone? Can the predictions be generalized or are they restricted to a particular pathogen-host system?

In general, what kinds of virulence factors (and how many, in relative terms) are sufficient in establishing virulence in a heterologous, a virulent organism?

What kinds of factors determine the evolutionary distance across which virulence might be genetically transferred? How does our current understanding of lateral gene transfer (its prevalence and role) in nature help us predict the degree to which virulence and other relevant phenotypes can be deliberately manipulated or created *de novo*?

What are the major challenges and unmet needs that hinder recognition and prediction of virulence? What would be the most effective strategies for addressing these challenges and needs?

What are the considerations for predicting function within systems?

What kinds of properties of a biological system are most successfully predicted from its genetic composition?

To what degree, and in what ways do current scientific knowledge allow construction of entirely new systems or organisms with predictable behaviors? How will this capability change as continued efforts increase our understanding of the function of individual proteins and macromolecular complexes?

What kinds of tools or approaches are currently most effective for predicting function from sequence? What are their capabilities? What are their limitations? How are they currently being used?

What are the major challenges and goals for developing more accurate predictive tools?

12:05 p.m. **Form and Context in Predicting Biological Function**

William Goldman, Ph.D.
Professor of Molecular Microbiology
Washington University School of Medicine

12:25 p.m. **Genotype to Phenotype**

Jim Musser, M.D., Ph.D.
Co-Director and Executive Vice President
The Fondren Foundation Distinguished Endowed Chair
The Methodist Hospital Research Institute

12:45 p.m. **Design Considerations for Robustness and Vulnerability in Biological Systems**

Marc W. Kirschner, Ph.D.
Chair, Department of Systems Biology Harvard Medical School

1:05 p.m. Design and Use of Predictive Tools: State of the Art

Owen White, Ph.D.
Director of Bioinformatics
Institute for Genome Sciences
University of Maryland School of Medicine

1:25 p.m. Discussion

2:45 p.m. *Break*

3:00 p.m. Session 3: Risk Assessment and Risk Management in a Context of Uncertainty

This session will explore the challenges of assessing biosafety risks in synthetic biology research when there is uncertainty about the biological properties of an agent, how biosafety risk assessment might be approached in such circumstances, and principles and strategies for risk management. Discussion questions are listed below.

Moderators:

Claudia Mickelson, Ph.D.
RAC Biosafety WG Member
Biosafety Program Deputy Director,
Environment, Health & Safety
Massachusetts Institute of Technology
Michael Imperiale, Ph.D.
NSABB Member
Professor
University of Michigan Medical School

Panelists:

Rocco Casagrande, Ph.D.
Managing Director
Gryphon Scientific
Lawrence McCray, Ph.D.
Research Associate,
Program on Emerging Technologies
Massachusetts Institute
of Technology

3:30 p.m. Discussion

Are there novel or distinct biosafety risks or challenges associated with synthetic biology? In general, to what degree are the biosafety risks of synthetic biology currently being adequately addressed?

For recombinant DNA research, the initial risk assessment is based on the risk group of the parental agent. However, for the more novel synthetic agents, a parental agent may not be obvious and/or the biological properties of the new organism may be largely unknown. How can risk assessment be conducted in such situations? What is the most appropriate risk management approach in such cases?

What kinds of efforts have been, or are being taken to engineer biological containment into synthetic systems/organisms (e.g, use of unnatural genetic code or amino acids, self-destruct mechanisms, other safeguards)? Are there other biological containment practices that could be considered?

Are there any existing risk assessment tools that would be applicable to the biosafety risk assessment process in the context of synthetic biology?

Once the difficult and inherently imprecise process of risk assessment has been completed and a risk management plan has been defined, are there ways to assess the efficacy of the management plan as the project goes forward?

4:50 p.m. Closing Remarks

5:00 p.m. *Adjourn*

Summary: Roundtable on Synthetic Biology
October 11, 2007
Convened by the National Science Advisory Board for Biosecurity
and the NIH Recombinant DNA Advisory Committee
Bethesda, Maryland

Welcome and Opening Remarks

Paul Keim, Ph.D., Northern Arizona University and the Translational Genomics Research Institute in Phoenix, Arizona, and voting member of the National Science Advisory Board for Biodefense (NSABB) committee, opened the roundtable, which was hosted by the NSABB committee and the NIH Recombinant DNA Advisory Committee (RAC).

Dr. Keim explained that the NSABB was established to recommend strategies for biosecurity oversight of dual use research, taking into consideration both national security concerns and the needs of the research community. The NSABB was given a two-part charge at its first meeting concerning synthetic genomics: to identify the potential biosecurity concerns raised by advanced nucleic acid synthesis technologies and to assess the adequacy of the current regulatory and oversight framework for addressing those issues. The NSABB has completed this phase of its charge and issued its findings and recommendations in its report Addressing Biosecurity Concerns Related to the Synthesis of Select Agents. The NSABB Working Group on Synthetic Genomes, which was formed to address the first part of the charge, is now poised to address the second phase—to identify and assess potential dual use concerns that may arise from work being performed in the broader field of synthetic biology. The experts gathered at this roundtable will share their knowledge and insights regarding the state of the science in synthetic biology. What is learned today and in future Working Group discussions of dual use potential with synthetic biology will be used to determine whether the field represents any novel biosecurity risks that would not be adequately addressed by the current oversight paradigm that the NSABB recently recommended.

Howard Federoff, M.D., Ph.D., Georgetown University Medical Center, and Chair, RAC, presented an overview and history of the RAC, from its establishment in 1974, in response to concerns raised by the advent of recombinant DNA, to today. He said that this roundtable is an important step in this tradition, because synthetic biology has advanced to the point where it soon will be able to generate novel biological entities that go beyond what can be achieved through traditional recombinant DNA approaches, and current guidance was developed with more traditional recombinant technology in mind. In light of new and advancing technologies, NIH has asked the RAC to revisit the concept of what a recombinant biological entity is and how to ensure that all relevant research is conducted safely. Toward this end, the RAC Biosafety Working Group will be exploring the current and future capabilities of synthetic biology and the implications for biosafety risk assessment and risk management. What is learned today will be applied in developing new draft principles and procedures for the safe conduct involving research using synthetic biological entities. Some of the questions NIH has asked the roundtable to explore include the following:

- What capabilities does synthetic biology provide beyond those achieved by recombinant DNA technology?
- Are there novel or distinct biosafety risks associated with synthetic biology?
- Currently, how are the biosafety risks of synthetic biology being addressed?
- What should be the principles of risk assessment and management for synthetic biology?
- Are the oversight systems in place for recombinant DNA research applicable to synthetic biology research?

Overview of Roundtable

David Relman, M.D., Stanford University, and Chair, NSABB Working Group on Synthetic Genomics, discussed the two phases identified as part of the charge to the group—the first involving issues related to the adequacy of the current regulatory framework for oversight on work involving select agents, and the second constituting a more encompassing set of tasks and issues related to whether there are dual use concerns that arise from synthetic biology, what they are, and how they can be addressed. The USG already has adopted a number of recommendations related to the issue of the current regulatory framework and select agents, among them one recommending that a system for screening sequences be examined and the features of such a system be fleshed out in some detail.

Dr. Relman outlined the day's goals as exploring the state of the science of synthetic biology and how it may be distinguished from other, related fields; considering whether one can predict the biological properties of a DNA construct from sequence and focusing on how we can assess and manage risk in this area. He outlined a number of difficult challenges, one of which is how to define the field in a practical, working sense.

Session 1: State of the Science of Synthetic Biology

Moderators:

Harvey Rubin, M.D., Ph.D.
NSABB Member
Director, Institute for Strategic Threat Analysis and Response
Professor of Medicine, Microbiology, and Computer Science
University of Pennsylvania

Stephen Dewhurst, Ph.D.
RAC Member
Professor of Microbiology and Immunology
and of Oncology University of Rochester

This session provided a broad understanding of the field of synthetic biology—the various research approaches, current capabilities, short and long-term goals.

A Bird's Eye View of Synthetic Biology
Roger Brent, Ph.D.,
Director and President,
The Molecular Sciences Institute

Dr. Brent presented an overview synthetic biology and the significance of the science involving the ability to predict biological function. He noted that biological function is not a term that is precisely defined, because it operates on multiple levels. Biological function can be explored from the use of high-throughout data, but this has been a slow process, and at the atomic level of function, DNA sequences are by far the best predictor of function. He emphasized that most of the gains in the understanding of individual gene function have come from the study of comparative genomics.

He discussed how systems biologists work, emphasizing that to understand and describe the quantitative behavior of any biological system, one must understand the key phenomena that describe its function and the physics that describes those phenomena. Only with these abstractions, which arise from close, thoughtful observation and consideration of living systems, can one achieve real quantitative understanding: without such abstractions, the work cannot succeed. This, Dr. Brent said, is how biology has been successful in the past. He discussed some scientific studies in this area, including some involving the study of yeast.

Dr. Brent also reviewed the history of recombinant DNA and regulatory schemes over the past 35 years, including the Asilomar model, and he emphasized some noteworthy points on the timeline, including 1994 and the advent of the web browser, the availability of the DNA sequence, and the origins of the term *synthetic biology*.

In the area of policy, Dr. Brent discussed DNA hacking and problems with some of the claims of novelty, because there are many more technical paths to engineering biological systems than are found in the synthetic biology canon. Thus the emphasis on synthetic biology may be distorting focus and causing the policy establishment to miss more important technical developments. He noted that there is no need to assemble long pieces of DNA in laboratories or by direct chemical synthesis; it may be easier and cheaper to do so in coli and in yeast.

He cautioned about the continued democratization and deskilling that is taking place in the hacking of DNA and organisms and about the entry of a new class of DNA hacker—someone who thinks of himself as an engineer of sorts first, rather than a scientist or researcher. He said that even though universities are providing first-rate ethical training to synthetic biologists, our society needs at all costs to avoid the creation and glamorization of such a class of outlaw hackers. Rather, we need to devise workable licensing schemes and a regulatory framework such as Asilomar, which has kept the peace since 1975.

Synthetic Biology as a 20-Year-Old Field

Steven A. Benner, Ph.D.
Distinguished Fellow
Foundation for Applied Molecular Evolution

Dr. Benner began by noting that synthetic genetic biology is nearly 25 years old, no risks have emerged, it underpins therapies that affect about 400,000 patients annually (e.g., with HIV, Hepatitis B and C), and is still constrained by the realities of organic chemistry. He added that there are many approaches to understanding biology—observation, analysis, reductionism, and synthesis. Synthesis guides discovery and innovation in ways that analysis cannot. With the other three approaches, if the data contradict theory, the tendency is to discard the data. In contrast, with synthesis, if at some point there is no agreement between theory and reality, one knows to stop—something is wrong and the investigator learns something about the system being constructed.

The power of chemistry comes from its ability to develop theory through synthesis of new matter. Several important molecular structures are associated with a major advance in structure theory in chemistry that came via synthesis, and which would not have occurred through analysis, for example development of the compound B12. Nonetheless, you cannot predict function based on structure—perhaps a naïve and overly optimistic hope of scientists involved in the Human Genome Project. Synthesis allows one to create structure to better understand function. The entry of synthesis into biology is a long tradition, dating back to biomimetic chemistry, the study of how enzymes work from serine proteases up to vitamin B12 using enzymes, recombinant DNA technology, and gene synthesis. Techniques such as codon optimization and strategic replacement of restriction sites are now routine in synthesis. In Benner's lab, synthesis has included total synthesis of genes, metabolic pathways, and bottom up synthesis of new proteins and new genetic systems.

Synthesis allows scientists to redesign a system to test theory against the reality of the model. Then one can determine whether the model, which is admittedly abstract, is able to support predictive chemistry. The question that challenges science, and philosophical debate—is whether one can or should create a self-sustaining artificial chemical system capable of Darwinian evolution. In addition, can these unnatural systems work in biological systems—which is the focus of “modern” synthetic biology. Benner stated that the future of synthetic biology is in self-avoiding genetic systems, i.e., DNA that binds to natural DNA, but not to other DNA. He also predicted that synthetic biology will produce: RNA molecules that catalyze template-directed RNA synthesis; understanding of how RNA emerged on prebiotic Earth; a 12-letter genetic system working in living cells; and a broad-based model of systems biology set in paleontological context.

He noted that biology has become more predictive by adopting chemistry’s meta-language. It is conceivable that this could allow us to predict virulence, for example. A challenge to the biological paradigm, however, is the notion that we could produce interchangeable parts in biological systems, a framework with which chemists are more familiar.

Dr. Benner concluded by stating that hazard in synthetic biology requires standard biochemistry (parasitism), self-sustenance, and the ability to evolve. In discussion, he agreed with Dr. Relman that it is not just hazards of unnatural origin that cause concern. Others raised the issue of inorganic-to-organic conversions and whether that changes the nature of what we call biology. In addition, one synthetic step can alter numerous other aspects in the organism that might not have been predicted. The more complicated the system the more likely it is that a single atomic perturbation will produce multiplexed, multiple interactions, most of which we do not understand.

Synthetic Biology: From Bacteria to Stem Cells

Ron Weiss, Ph.D.
Professor of Electrical Engineering
Princeton University

Dr. Weiss related that he was a computer scientist by training and became fascinated with the notion that we might be able to program cells with the same ease and capability that we program computers. He noted that synthetic biology uses genetic engineering principles and techniques to figure out ways to design complex systems. To design DNA, we need the ability to synthesize long pieces of DNA very quickly, which requires the ability to understand the system and knowledge of a mechanism by which to take parts in a rational computer-assisted way and put them together to achieve a predetermined purpose. Existing tools facilitate the analysis of systems with hundreds or thousands of components but are not useful for design.

A bottom-up approach assumes, for example, that if you want to have a particular biological property embedded in a biological system, you can take an existing biological system that exhibits behavior that is somewhat close to what you are seeking, and through some mechanism (e.g., directed evolution or cross breeding), get closer to the function that you want. A bottom up approach would be similar to the development of software, in which several versions are tested before arriving at a successful one. Biological systems, however, unlike computer systems are not predictable. In addition, several issues in engineering biology have to be considered, such as:

- Device characterization
- Rules of composition
- Noise

- Cellular context
- Mutations
- Environmental conditions
- Rational design vs. directed evolution
- Crosstalk
- Impedance matching
- Cell death
- Chemical diffusion
- Motility
- Reliance on incomplete models

The systems do not have to be designed in the same way as evolution. The beauty of synthetic biology is that you one can make intermediate systems that are not very functional or highly optimized to work within a particular context. Those might allow us to think about how to make version 2.0 or version 3.0. The goal of synthetic biology is to set the foundation for building modules—for example, cascades, toggle switches, pulse generators, ultrasensitive switches, or oscillators. One can then modify certain attributes of the modules to imagine how the system would then be modified. Computational tools will be needed to understand how the modules will interact within the system, relying on digital logic to understand the potential cascade of events and why certain designs do or do not work.

Dr. Weiss provided an example of applying synthetic biology to stem cell research, which poses a fundamental question in tissue engineering—can we create large scale spatially predefined tissue patterns? He stated that we can use our experience with bacterial synthetic multicellular systems to implement sophisticated rules of interactions between mammalian stem cells that result in spatial patterns of differentiation. The goal would be programmed tissue (re)generation through differentiating stem cells in space and time into desired 3-D patterns. Once the system is understood, the challenges is making it work *in vitro.* The concepts of design and programming could be extended to engineering mammalian cells to communicate with one another or to program multiple steps of interaction. He cited diabetes research in which a genetic network could be designed to drive beta cell development.

Dr. Weiss concluded by emphasizing that what synthetic biology needs is design principles, i.e., engineering rules. The construction of basic modules can then be tested in applications such as programmed tissue regeneration, artificial tissue homeostasis, and artificial immune systems.

Synthetic Organisms
Steen Rasmussen, Ph.D.
Team Leader for Self-Organizing Systems
Los Alamos National Laboratory

The "bottom-up" approach to synthetic biology attempts to assemble minimal living systems by taking nonliving components and putting them together in a variety of ways.

The premise, said Dr. Rasmussen, is if we understand how to make living systems from scratch, we can probably make technology based on the same principles as living systems. That capability would be robust and autonomous, and have local intelligence and the ability to repair itself and evolve. Dr. Rasmussen discussed the difficulty in defining "minimal life," which is notoriously complex.

One operational definition is based on three interconnected functionalities that can transform resources into building blocks that grow, divide, and undergo evolution.

Because building protein synthesis machinery is incredibly complicated, it can be sidestepped by designing a metabolic system where the efficiency is determined by the sequence; thus, if you can replicate the sequence you have a hope for selection because bad metabolism/good metabolism means that you can select the best outcome.

Rasmussen's work focuses on vesicles—how they grow on a particular structure and how it grows. He is focused on representing, generating, analyzing, and controlling self-organizing and related systemic processes as they are manifested in natural and human-made systems.

Dr. Rasmussen described his work involving assembly of protocells. He emphasized that the question is not whether new simple life forms can be assembled, but under which conditions they can be assembled.

He believes that eventually we can use this technology to build protocells that will be increasingly autonomous and able to be weaned off microfluidics support. This "living technology" could have a large socioeconomic impact in 20-25 years.

But it will only be realized through a deep understanding of the nature of living processes, which can occur through making life from scratch. If successful, this approach could have applications in the development of self-healing materials, medical diagnostics and treatment, security (the ability to recognize and neutralize bioagents, or modify chemical composition of nuclear waste), environmental protection, and energy production.

Discussion

One of the biggest challenges for synthetic biology—as is the case for any new or evolving field—is arriving at a definition that everyone can agree on. This has implications for oversight and policy. For example, for oversight purposes, if synthetic biology includes designing an organism that is a potentially self-replicating, evolving entity with predictable properties and an anticipated evolutionary or adaptive rate, does that require oversight? Many participants felt that a critical consideration is whether the new entity is self-sustaining.

In fact, the goal of some current research is design for self-replication. The real challenge is building a system that is able to exhibit open-ended evolution. The view from the bottom-up experts was that the science is still very far away from achieving the goal of self-replication.

One more pragmatic goal would be not to predict function of a completely random sequence of DNA but rather to focus on being able to predict functions of a restricted set of DNA sequences. One might then be able to predict the behavior of a single base mutation or at least a small set of mutations from that given set of initial sequences. Function can be defined on many levels, however, a certain level of functional insight is sufficient to predict a given outcome.

Thus, both bottom-up and top-down approaches are needed. There was some agreement that the more immediate concerns will arise in the top-down approaches, e.g., modifying mycoplasma through knowledge of certain polymorphisms and their significance.

There was some discussion of building in safety features to any design based on the statistical probability that a given failure will occur. However, biological systems work on different principles than do engineered or computational systems, for example, there is variation in selection, some of which is random and some of which is adaptive.

What is needed is a biological programming language that includes all of the underlying features of biological systems.

The discussion ended with broad agreement that although there are no imminent risks raised by synthetic biology, the public, and even parts of the scientific community are misinformed about the goals and limitations of the field; thus, education is essential.

Session 2: Predicting Function

This session addressed current understanding of the relationship between biological properties and sequence and structure, our ability to predict biological function/properties, and the tools that are available or under development for predicting function.

Moderators:

Claire Fraser-Liggett, Ph.D.
NSABB Member
Director, Institute of Genome Sciences
University of Maryland School of Medicine, Baltimore

Form and Context in Predicting Biological Function
William Goldman, Ph.D.
Professor of Molecular Microbiology
Washington University School of Medicine

Nikunj Somia, Ph.D.
RAC Member
Assistant Professor
University of Minnesota, Twin Cities

As a microbiologist, Dr. Goldman studies pathogens, trying to define what is required for virulence based on sequence—more of an analytical and reductionist approach than a synthetic one. Bacteria are the perfect example of how "form follows function" because the external design of these rather simple organisms reveals the exact architecture of the rigid peptidoglycan skeleton underneath.

Dr. Goldman described a study with *Bordetella pertussis*, the whooping cough agent, where a specific peptidoglycan fragment called "tracheal cytotoxin" was found to be responsible for much of the pathology in the disease. The current model of how this works is that the organisms attached to the ciliated cells in the respiratory tract release tracheal cytotoxin, along with endotoxin (LPS), and that triggers a series of events inside the neighboring cells resulting in production of a large amount of nitric oxide. It is the host cell production of nitric oxide that kills off the ciliated cells and forces their ejection from the epithelium. A few other organisms release little pieces of peptidoglycan for specific purposes, one of which is *Vibrio fischeri*; in this case, the peptidoglycan is important for light organ development in the Hawaiian bobtail squid, which demonstrates a symbiotic use of the same molecule. The significance of this is that a virulence factor in one system is not always a virulence factor in another—it is a matter of host interpretation—and that sometimes function follows form. The behavior of a molecule can be context-dependent, and sequence information will not tell you that.

Another example is *Histoplasma capsulatum*, a fungal pathogen that also causes a respiratory tract disease. It is a dimorphic fungus, existing as either a mold or yeast. It can be switched from mold to yeast just by raising the temperature *in vitro* to 37 degrees celsius. Thus, it changes lifestyle when encountering a mammalian host. The best-studied *Histoplasma* virulence factor is a small yeast phase-specific protein called CBP, and no hints about its function have come from sequence homologs or motifs. However, the 3-dimensional structure of CBP has structural homologs that provide major clues regarding function, even though the primary amino acid sequence did not. In this case, as with *Bordetella pertussis*, sometimes function follows form, and the only way to get at function is to actually look closely at the molecule and do the biochemistry.

Genotype to Phenotype

Jim Musser, M.D., Ph.D.
Co-Director and Executive Vice President
The Fondren Foundation Distinguished Endowed Chair
The Methodist Hospital Research Institute

Dr. Musser described his research involving molecular dissection of epidemic waves and strain genotype-infection phenotype in Group A *streptococcus*, the flesh eating bacteria. One ongoing project is a study of the molecular genetic processes contributing to epidemics and clone emergence (using serotype M1 and M3 strains). Dr. Musser's lab is attempting to understand to what extent they can get to a predictive model of epidemics and clone emergence. They also are attempting to develop a predictive model of what at the genetic level mediates disease specificity. These approaches involve an integrated strategy for studying bacterial pathogenesis, including genome sequencing, development of infection models in the mouse and nonhuman primates, iterative expression microarray analysis, bioinformatics, and human specimens and accompanying clinical data. To date they have 12 Group A strains chosen for their probe-specific genotype-phenotype relationships, e.g., extremely high virulence or associated with post infectious sequelae like acute glomerulonephritis and acute rheumatic fever.

Dr. Musser's lab has learned that acquisition of bacteriophages expressing novel virulence factors is a crucial issue in clone emergence and disease specificity. Permutation of virulence factors is also an important effect, i.e., strep permutates its genome with mobile elements. In addition, genetic inactivation of one particular gene that results in up-regulation of virulence factors can be important in clone emergence and disease specificity. There are multiple permutations in Group A occurring during its daily activities in the human host. This is all complicated by the fact that Group A streptococci differ at up to 15 percent in chromosomal gene content. Thus, a sequence-based predictive model of behavior is currently not possible.

Other work involves developing a reasonable model system in the human to understand what molecular forces may be contributing to phenotypically distinct epidemic waves. Each distinct epidemic wave has very distinct nonrandom phenotypic traits. However, one clone caused significantly fewer cases of necrotizing fasciitis despite it being the most common genotype. It was subsequently found that a truncation mutation in the mtsR gene—thought to be important to growth and virulence *in vivo*—may have been the factor lowering the virulence of that clone. Dr. Musser noted that numerically speaking, most of the events that differentiate one strain of bacteria from another are single nucleotide polymorphisms. They are modest changes in the genome that contribute significantly to distinct disease specificity. It is critical in moving forward to have an integrative investigative approach in which strains carefully matched with patient phenotype are used. It is likely that in many infectious agents rare alleles are very important mediators of disease phenotype.

Design Considerations for Robustness and Vulnerability in Biological Systems

Marc W. Kirschner, Ph.D.
Chair, Department of Systems Biology
Harvard Medical School

Dr. Kirschner began by agreeing with previous speakers that you cannot really understand a biological entity until you can make it from scratch, a concept that chemists have long embraced. As for prediction, there are three major types of information to predict things from—structure, genes, and databases. Accurate prediction could short circuit very difficult experiments in toxicology and drug design, in terms of both efficacy and of side effects.

It is difficult in many system to predict function because the number of genes in a complex animal is surprisingly low; the number of types of signaling pathways is exceedingly low; and pathways adapt genetically and physiologically. There is also an unexpected paradox of conservation which makes it difficult to explain diversity. The groundbreaking work of Beadle and Tatum did not consider the role of context. Thus, their brilliant effort at the first genotype-phenotype map turned out to be overly simplistic and unworkable for multicellular organisms.

Systems biology aims to understand the versatility of these conserved core processes for the purpose of predicting function, which is very context dependent. Interestingly, it turns out that one of the features of biological system is that the components do not change all that much, which is why there are so few genes. What does change is regulation. Systems biology needs to understand not only the structure of processes in terms of current use but also their modifiability in evolution. It needs to address on a higher level the robustness of processes; this can only be done on a level where the tradeoffs between constraint and deconstraint can be evaluated. It also needs to understand what is accessible in evolution—the range of each process must be considered not merely the range of the organism. Thus, we need to understand the adaptive nature of the engineered organisms and also the adaptive nature of the hosts that have not been engineered. Dr. Kirschner provided a detailed example of Wnt signaling as a real life circuit, whose structure is just being understood. It raises questions as to why conserved pathways do not change— perhaps form does follow function. He concluded that we know very little about predicting function. The goal of synthetic biology is a more predictive and ultimately quantitative relationship of genotype to phenotype. Even for microorganisms with relatively small genomes, we have insufficient knowledge to make quantitative predictions. Though we can show genetic requirements, it is much harder to predict fitness for systems that operate far from steady state. Developments in this field should aid in the production of new drugs and in predicting the behavior of organisms in novel hosts.

Design and Use of Predictive Tools: State of the Art

Owen White, Ph.D.
Director of Bioinformatics
Institute for Genome Sciences
University of Maryland School of Medicine

Dr. White began by emphasizing that there is a range of risk in synthetic biology, with the bottom-up approaches— entirely synthesized cells—on one end the spectrum and modified living cells on the other end. He then proceeded to talk about the design and use of predictive tools, specifically, to what degree are we able to explain virulence of the more common bacterial and viral pathogens, based on their genome sequences?

He described a "genome property" as "an attribute of biological organisms that is rigorously defined such that assertion of its absence, presence, or quantitative extent can be made (either automatically or manually) in a self-consistent manner." The property could be any type of biological processes, including metabolic pathways, observable phenotypes, and quantitative measures of genomic content. Dr. White's lab employs a pattern recognition method to develop probability tables to determine how closely an unknown sequence matches known sequences (i.e., to determine conserved proteins). Hidden Markov Models (HMMs) allow automated assignment of sequences to homology families. The purpose is to detect families having the same function based on conserved peptide positions.

The properties of a biological system most successfully predicted from its genetic composition include: amino acid metabolism, polyketide and non-ribosomal peptides, and cofactor and vitamin metabolism. The ability to correctly predict pathways was successfully tested by running a new genome sequence for which sequencing had not yet been done. Other investigators are attempting similar predictions based on different approaches, for example, Ross Overbeek has developed something called "Subsystem," and Eugene Koonin's clusters of autologous groups.

The next questions are: How accurately can virulence or other pathogenic properties be predicted on the basis of sequence alone? Can the predictions be generalized or are they restricted to a particular pathogen-host system? Advances in this area have been accelerated through bioinformatics and resource centers funded by NIH. The focus of some of these activities is curating genomic sequences of pathogens and looking at different strains to identify products that might become vaccines, therapeutics, or diagnostics. The types of data being gathered could be rolled into a genome property system that could be used to predict virulence of new bacterial genome. As for predicting function from sequence, there is sufficient evidence that regulation might be as important, if not more so, than sequence.

Discussion

The fact that genes code for proteins is but a small piece of the puzzle in predicting function.

With regard to synthesizing a virulent pathogen, there was skepticism that all prediction of function could occur anytime soon since no one has figured out how to synthesize something that self replicates. If someone wanted to act maliciously, it would be far more expedient to modify a known pathogen, e.g., smallpox, to enhance virulence (the "top-down" approach).

Loss of function seems to be a central featured of increased virulence, thus, that would be a logical phenomenon to try to understand. Another approach would be to modify the host in some way. There was some discussion about how to characterize risk, especially for dual use research.

A few simple rules for responsible research have already been in place in this area of research.

For example, if you want to investigate self-reproducing programs and spread of those inside a computer, you have to simulate a computer inside a computer, that is, you cannot use the operation system itself. History has shown that it is difficult to predict the scenario in which an accident might happen because often it is a series of two or three sequential events. It is also difficult to regulate in a uniform way because the pathogens are so unique, requiring cell and animals models for study.

What makes virulence so complex is that much of infectious disease pathology is driven by the immune response of the host, which is highly variable and dependent on immunogenetics.

There was agreement that understanding specific organisms is incredibly complicated by their biological context.

Session 3: Risk Assessment and Risk Management in a Context of Uncertainty

This session explored the challenges of assessing biosafety risks in synthetic biology research when there is uncertainty about the biological properties of an agent, how biosafety risk assessment might be approached in such circumstances, and principles and strategies for risk management.

Moderators:

Claudia Mickelson, Ph.D. RAC
Biosafety WG Member Biosafety Program
Deputy Director, Environment, Health & S
afety Massachusetts Institute of Technology

Michael Imperiale, Ph.D.
NSABB Member
Professor
University of Michigan Medical School

Dr. Imperiale began the session by saying that judgments need to be made regarding what types of risks certain synthetic biological experiments might pose and how they should be managed. He reminded participants that this discussion would go beyond covering human pathogens to elements that affect agriculture, animals, and the environment. Dr. Mickelson added that it is important when communicating to the public about this science to use language that is more descriptive and realistic about its goals. She emphasized the importance of having strong advocates to communicate with the public using paradigms that are easily explained in order to convey the seriousness with which scientists regard this research and their sense of responsibility regarding any risks that may be involved. She said she looked forward to hearing panelists' thoughts about how relevant and reasonable criteria can be developed to assess this research so that it moves forward as rapidly as possible.

Panelists:

Rocco Casagrande, Ph.D.
Managing Director Gryphon Scientific

Lawrence McCray, Ph.D.
Research Associate, Program on Emerging
Technologies Massachusetts Institute of Technology

Dr. Casagrande focused his comments and observations on risk management and risk assessment as they relate to completely synthetic organisms, suggesting that such observations might help inform biosafety and biosecurity guidelines. He noted that scientists known as bioprospectors bring organisms with which we have no previous experience into the human realm every day, but these scientists are not required to wear extensive protective equipment, largely because they claim that there is no selective pressure on these organisms to be pathogenic. This same argument could be made for the results of DNA synthesis. He also discussed how hospitals and hospital laboratories handle risk from unknown microbes routinely without imposing extreme levels of bioprotection or isolation. Although there always is some risk that someone in a hospital may be infected with a pathogen that requires a higher level of containment, the cost of broad containment is so high that it would cause our infectious disease system to grind to a halt. He also emphasized the astounding diversity of the microbial world and yet the not-so-astounding diversity of emerging pathogens, which generally fall into predictable categories. And, he noted, even when a new virus such as SARS emerges, it is usually related to an existing animal pathogen, because pathogens that have experience with humans already know how to evade the human immune system and elbow out competing microbes. Something completely synthetic that has no experience with humans has a very little chance of doing that.

Dr. Casagrande suggested that it may be advisable to control selective pressure instead of trying to regulate the creation of organisms and that when adding pathogenic components to nonpathogenic organisms, the organism should be treated at the higher BSL level of the pathogen until information is available that indicates otherwise. He suggested that an important question to ask is whether the researcher or the PI is the right person to make the decision about what is safe. He also said it would be important to ask if a higher level of scrutiny is needed when dealing with known pathogenic elements in nonpathogenic organisms.

Dr. McCray drew on his Washington policy experience at the NAS and in the Executive Office of the President to lay out three questions that the synthetic biology community may want give greater consideration.

1) From a policy perspective, what is really new about synthetic biology; does it really present a new type of risk beyond what safeguards contemplate? In asking MIT researchers about this, he mentioned that one person responded that the risk wasn't new, but the reduced costs of doing synthetic biology may mean that new classes of people—well beyond academic researchers —may have access to it, and that today's safeguards will not automatically reach those people.
2) Traditionally, those who are worried about the risks of emerging technologies have put their main effort into prevention. The record of experience, however, suggests that past technological predictions are rarely reliable. Can the research community help find adaptive mechanisms that will detect and respond to any future unexpected adverse effects of synthetic biology?
3) Several attendees today have expressed concern about future "garage-level" work on synthetic biology— analogous to the garage software shops that are found in the "dot.com" sector. Is this a real possibility, and if so, on what time scale? Can the research community provide input on this question?

Discussion

Dr. Imperiale said that that both the NSABB and the RAC would be interested in hearing about whether there are new or different types of risks that are presented by synthetic biology. If it is believed that there are no new risks and that we can use an existing oversight structure, the situation would be very different from one in which new risks are apparent that will require new means of assessment and management.

There was agreement that what is new in the area of synthetic biology is that the cost is becoming lower and the technology is disseminating and that this is making it easier for hackers to enter the field. It would be difficult to stop them, because at any given time, a half million people have access to "how-to-clone" cookbooks in the United States alone. Some argued that the question is whether the Asilomar paradigm will survive when it becomes possible for those who are not biologists or university faculty to work on their own. It was suggested that the Asilomar paradigm could continue to work, so long as everyone who has access to the technology subscribes to it.

The risk of acquisition by adversaries was discussed, as were possible risks involving media reports about a stolen organism that could cause panic and result in severe economic consequences.

It also was suggested that what is new is the ever-increasing enabling technologies of high-throughput sequencing and synthesis, which together create a new capability. The question is whether a new system of oversight is needed and whether our university systems are prepared to deal with these technologies and their possibilities, such as making recombinant retroviruses that can infect human cells.

In addition, it was pointed out that recently the House Energy Commerce Subcommittee held a hearing about the biosafety of high containment laboratories and that GAO questioned whether there has been adequate biosafety supervision and training—issues the scientific community must address.

In summarizing comments on this subject to this point, Dr. Federoff noted that there has been a confluence of new approaches that enable biology to move much faster and that will spawn may new insights that in time will be highly relevant to what the RAC currently considers.

These approaches may be of value to us societally, but they also may carry associated risk and will require a high degree of vigilance. As the evidence comes forward, it needs to be evaluated prospectively and it needs to be assessed in the context of whether this represents an apparent risk or a real risk, and if it is a real risk, it needs to be attended to in a formal way.

He also outlined how the two subcommittees and ultimately the parent advisory committee bodies might be able to constructively interact and said that more could be accomplished at this gathering to codify an effort involving risk evaluation and mitigation that is based on biological context. The items that warrant the most discussion are those that lie in the gray area regarding their purview under the RAC, and it is the biological context that makes them relevant.

Further discussion involved how those present have been dealing with biosafety as it relates to the kind of work presented in the morning sessions. Most reported a perception of no or low risk in their activities, and participants described some of the systems used for analyzing and understanding risk.

One comment was that although the current NIH Guidelines speak primarily to traditional recombinant techniques, they also speak to synthetic DNA, but not unequivocally.

It was suggested that developments in the field of risk assessment be monitored, and there was some discussion of broadening the charge to include the synthesis of infectious agents not just by infecting existing replicating cells but also by using in vitro methods. Participants agreed that most of what is under discussion already falls under existing guidelines, but that there are concerns involved in bridging the line between chemistry and life.

Dr. Weiss ended the session by proposing that it may be useful to envision a spectrum where on one end there are the existing definitions perhaps by the RAC that genes and organisms define risk and on the other end there are random DNA sequences for which we have no way of predicting what may happen.

Somewhere in the middle may be the notion that we are trying to create new DNA sequences that are sufficiently different from what exists now. Because it would be difficult to establish well-specified guidelines for dealing with this middle area, it might be best in such cases to have the investigator who is familiar with a particular project conduct a risk assessment.

Closing Remarks

Dr. Imperiale and Dr. Relman thanked all participants for their contributions and emphasized the importance of involving the public in continued discussions.

End Notes

[1] National Science Advisory Board for Biosecurity, Proposed Framework for the Oversight of Dual Use Life Sciences Research: Strategies for Minimizing the Potential Misuse of Research Information (Washington, DC: June 2007), http://oba.od.nih.gov/biosecurity/ biosecurity_documents.html.

[2] Drew Endy, "Foundations for engineering biology," Nature 438 (2005): 449-453.

[3] Meeting of the National Science Advisory Board for Biosecurity, June 30 – July 1, 2005 http://oba.od.nih.gov/biosecurity/nsabb_past_meetings.html.

[4] Select Agents are biological agents and toxins regulated by the Select Agent Rules (7 CFR Part 331, 9 CFR Part 121, and 42 CFR Part 73) that have the potential to pose a severe threat to public, animal or plant health, or to animal or plant products. A list of Select Agents and Toxins can be found at http://www.selectagents.gov/ Select%20Agents%20 and%20 Toxins%20List.html (accessed 10/29/09).

[5] National Science Advisory Board for Biosecurity, Addressing Biosecurity Concerns Related to the Synthesis of Select Agents (Washington, DC: December 2006), http://oba.od.nih.gov/ biosecurity/biosecurity_ documents. html.

[6] National Science Advisory Board for Biosecurity, Proposed Framework for the Oversight of Dual Use Life Sciences Research: Strategies for Minimizing the Potential Misuse of Research Information (Washington, DC: June 2007), http://oba.od.nih.gov/biosecurity/ biosecurity_documents.html.

[7] http://syntheticbiology.org/ (accessed 9/29/09).

[8] Ibid.

[9] Steen Rasmussen et al., "Evolution: Transitions from Nonliving to Living Matter," Science 303, no. 5660 (2004): 963-965.

[10] "Self-referential" in this context refers to biological entities or systems that are composed of elements that can interact, respond to changes, self-modify, replicate or expand.

[11] Travis S. Bayer et al., "Synthesis of Methyl Halides from Biomass Using Engineered Microbes," J. Am. Chem. Soc. 131 (2009): 6508–6515.

[12] Andres Moya et al., "Toward minimal bacterial cells: evolution vs. design," FEMS Microbiol. Rev. 33, (2009): 225-235.

[13] Daniel G. Gibson et al., "Complete chemical synthesis, assembly, and cloning of a Mycoplasma genitalium genome," Science 319, no. 5867 (2008): 1215-1220.

[14] John I. Glass et al., "Essential genes of a minimal bacterium," Proc. Natl. Acad. Sci. 103, no. 2 (2006): 425-430.

[15] C. Lartigue et al., "Creating Bacterial Strains from Genomes That Have Been Cloned and Engineered in Yeast," Science 325, no. 5948 (2009): 1693-1696.

[16] The BioBricks Foundation; http://bbf.openwetware.org/ (accessed 10/1/09).

[17] http://syntheticbiology.org (accessed 9/29/09).

[18] Drew Endy, "Foundations for engineering biology," Nature 438 (2005): 449-453.

[19] Ibid.

[20] Ravi Singh and Kostas Kostarelos., "Designer adenoviruses for nanomedicine and nanodiagnostics," Trends in Biotechnology 27, no. 4 (2009): 220-229.

[21] Robert D. Barish et al., "An information-bearing seed for nucleating algorithmic self-assembly," Proc. Natl. Acad. Sci. 106 (2009) 6054-6059.

[22] Pengcheng Fu, "Biomolecular computing: Is it ready for take off?" Biotechnology Journal 2 (2007): 91-101.

[23] Meeting of the National Science Advisory Board for Biosecurity, October 11, 2007, http://oba.od.nih.gov/biosecurity/nsabb_past_meetings.html.

[24] Recombinant DNA Advisory Committee website: http://oba.od.nih.gov/rdna/rdna.html.

[25] National Science Advisory Board for Biosecurity, Addressing Biosecurity Concerns Related to the Synthesis of Select Agents (Washington, DC: December 2006), http://oba.od.nih. gov/biosecurity/biosecurity_ documents.html.

[26] Juli Peretó and Jesús Català, "The renaissance of synthetic biology," Biological Theory, 2, no. 2 (2007): 128-130.

[27] http://www.ncbi.nlm.nih.gov/genomes/lproks.cgi?view=1 (accessed 10/7/09).

[28] Maung Nyan Win et al., "Frameworks for programming biological function through RNA parts and devices," Chemistry and Biology 16 (2009): 298-310.

[29] Jesse W.-H. Li and John C. Vederas, "Drug discovery and natural products: End of an era or an endless frontier?" Science 325 no. 5937 (2009): 161-165.

[30] Michael A. Fischbach and Christopher T. Walsh, "Antibiotics for emerging pathogens," Science 325 no. 5944 (2009): 1089-1093.

[31] Samuel K. Sia et al., "Synthetic tissue biology: Tissue engineering meets synthetic biology," Birth Defects Research 81 (2008): 354-361.

[32] M. P. Lutolf and J. A. Hubbell, "Synthetic biomaterials as instructive extracellular microenvironments for morphogenesis in tissue engineering," Nature Biotechnology 23 (2005): 47-55.

[33] Stuart Kyle et al., "Production of self-assembling biomaterials for tissue engineering," Trends in Biotechnology 27, issue 7 (2009): 423-433.

[34] Stephen Picataggio, "Potential impact of synthetic biology on the development of microbial systems for the production of renewable fuels and chemicals," Curr. Opin. Biotechnol. 20, issue 3 (2009): 325-329.

[35] Sung Kuk Lee et al., "Metabolic engineering of microorganisms for biofuels production: from bugs to synthetic biology to fuels," Curr. Opin. Biotechnol. 19 issue 6 (2008): 556-563.

[36] Travis S. Bayer et al., "Synthesis of Methyl Halides from Biomass Using Engineered Microbes," J. Am. Chem. Soc. 131 (2009): 6508–6515.

[37] Victor de Lorenzo, "Systems biology approaches to bioremediation," Curr. Opin. Biotechnol. 19 issue 6 (2008): 579-589.

[38] Elizabeth L. Rylott and Neil C. Bruce, "Plants disarm soil: engineering plants for the phytoremediation of explosives," Trends in Biotechnology 27 issue 2 (2009): 73-81.

[39] National Science Advisory Board for Biosecurity, Addressing Biosecurity Concerns Related to the Synthesis of Select Agents (Washington, DC: December 2006), http://oba.od.nih.gov/biosecurity/biosecurity_ documents.html.

[40] http://www8.nationalacademies.org/cp/projectview.aspx?key=49063 (accessed 10/16/09).

[41] David S. Kong et al., "Parallel gene synthesis in a microfluidic device," Nuc. Acids Res. 35 no. 8 (2007): e61.

[42] Harris H. Wang et al., "Programming cells by multiplex genome engineering and accelerated evolution," Nature 460 (2009): 894-898.

[43] Keith E.J. Tyo et al., "Stabilized gene duplication enables long-term selection-free heterologous pathway expression," Nature Biotechnology 27 no. 8 (2009): 760-765.

[44] Joseph R. Warner et al., "Genomics enabled approaches in strain engineering," Curr. Opin. Microbiol. 12, no. 3 (2009): 223-230.

[45] The NIH Guidelines and more information about the Recombinant DNA Program, the RAC, and IBCs are available at http://oba.od.nih.gov/rdna/rdna.html.

[46] Federal Register notice of consideration of a proposed action under the NIH Guidelines: http://oba.od.nih.gov/oba/RAC/meetings/jun2009/Final%20Published%20FRN.pdf.

[47] National Science Advisory Board for Biosecurity, Proposed Framework for the Oversight of Dual Use Life Sciences Research: Strategies for Minimizing the Potential Misuse of Research Information (Washington, DC: June 2007), http://oba.od.nih.gov/ biosecurity/ biosecurity_documents.html.

[48] National Science Advisory Board for Biosecurity, Strategic Plan for Outreach and Education on Dual Use Research Issues (Washington, DC: December 2008), http://oba.od.nih.gov/ biosecurity/biosecurity_ documents.html.

[49] Drew Endy, "Foundations for engineering biology," Nature 438 (2005): 449-453.

[50] J.H. Kuhn. "Filoviruses, A Compendium of 40 Years of Epidemiological, Clinical, and Laboratory Studies," SpringerWienNew York 2008: 37-58.

[51] Maung Nyan Win et al., "Frameworks for programming biological function through RNA parts and devices," Chemistry and Biology 16 (2009): 298-310.

[52] Jeffrey J. Tabor et al., "A Synthetic Genetic Edge Detection Program," Cell 137 (2009): 1272–1281.

[53] BioBricks Foundation: http://bbf.openwetware.org/ (accessed 10/1/09).

[54] National Science Advisory Board for Biosecurity, Proposed Framework for the Oversight of Dual Use Life Sciences Research: Strategies for Minimizing the Potential Misuse of Research Information (Washington, DC: June 2007), http://oba.od.nih.gov/biosecurity/ biosecurity_documents.html.

[55] iGEM: http://2009.igem.org/Main_Page.

[56] DIYbio: http://diybio.org (accessed 10/1/09).

In: Synthetic Biology
Editors: Roger E. Hamilton and Laura M. Morales
ISBN: 978-1-62257-207-6

Chapter 3

STATEMENT OF J. CRAIG VENTER, PRESIDENT, J. CRAIG VENTER INSTITUTE. HEARING ON "EFFECTS OF DEVELOPMENTS IN SYNTHETIC GENOMICS"*

Mr. Chairman and Committee members, I welcome the opportunity to testify before you today. I am J. Craig Venter, Ph.D, President and Founder of the J. Craig Venter Institute (JCVI). The JCVI is a not-for-profit research institute in Rockville, MD and La Jolla, CA dedicated to the advancement of the science of genomics; the understanding of its implications for society; and communication of those results to the scientific community, the public, and policymakers. The JCVI is home to approximately 400 scientists and staff with expertise in human and evolutionary biology, genetics, bioinformatics/ informatics, information technology, high-throughput DNA sequencing, genomic and environmental policy research, and public education in science and science policy. The JCVI is a 501 (c) (3) organization. I am also the Founder and Chief Executive Officer of Synthetic Genomics Incorporated (SGI) a privately held company launched in 2005 to speed the commercialization of synthetic genomics technologies for a wide variety of applications including energy, the environment, and medicine. In my testimony today I will first provide a brief overview of synthetic genomics, including answers to some key questions that I am often asked about this new technology. I will then briefly describe our recent accomplishment, and the 15 years of research that preceded it. Finally, I will discuss work to date on the ethical and societal implications of synthetic biology and review the ongoing policy discussions within the Federal Government.

OVERVIEW

Genomic science has greatly enhanced our understanding of the biological world. It is enabling researchers to "read" the genetic code of organisms from all branches of life by determining the sequence of the four letters that make up DNA. Sequencing genomes has now become routine, giving rise to thousands of genomes in the public databases. In essence, scientists are digitizing biology by converting the A, C, T, and G's of the chemical makeup of

* This is an edited, reformatted and augmented version of testimony given on May 27, 2010 before the House Committee on Energy and Commerce.

DNA into 1's and 0's in a computer. But can one reverse the process and start with 1's and 0's in a computer to define the characteristics of a living cell? We set out to answer this question.

In the field of chemistry, once the structure of a new chemical compound is determined by chemists, the next critical step is to attempt to synthesize the chemical. This would prove that the synthetic structure had the same function of the starting material. Until now, this has not been possible in the field of genomics. Structures have been determined by reading the genetic code, but they have never been verified by independent synthesis. In 2003, JCVI successfully synthesized a small virus, approximately six thousand base pairs long, that infects bacteria. And by 2008, the JCVI team was able to synthesize a small bacterial genome, 580,000 base pairs long. My team and I have now achieved the final step in our quest to construct the first synthetic bacterial cell. In a publication in *Science* magazine, Daniel Gibson, Ph.D. and a team of 23 additional researchers outline the steps to synthesize a 1.08 million base pair *Mycoplasma mycoides* genome, constructed from four bottles of chemicals that make up DNA. This synthetic genome has been "booted up" in a cell to create the first cell controlled completely by a synthetic genome. The work to create the first synthetic bacterial cell was not easy, and took this team approximately 15 years to complete. Along the way we had to develop new tools and techniques to construct large segments of genetic code, and learn how to transplant genomes to convert one species to another. The 1.08 million base pair synthetic *M. mycoides* genome is the largest chemically defined structure ever synthesized in the laboratory. While this first construct—dubbed *M. mycoides* JCVI-syn1.0—is a proof of concept, the tools and technologies developed to create this cell hold great promise for application in many critical areas. Throughout the course of this work, the team also contemplated, discussed, and engaged in outside review of the ethical and societal implications of their work. The ability to routinely write the "software of life" will usher in a new era in science, and with it, new products and applications such as advanced biofuels, clean water technology, food products, and new vaccines and medicines. The field is already having an impact in some of these areas and will continue to do so as long as this powerful new area of science is used wisely. Continued and intensive review and dialogue with all areas of society, from Congress to bioethicists to laypeople, is necessary for this field to prosper.

Answers to Some Key Questions

I would like to give you an overview of the potential for the new field of synthetic genomics and the implications of our work to construct a synthetic cell by providing brief answers to a series of key questions.

How Is Synthetic Genomics Different Than Standard Molecular Biology/Genetic Engineering, etc?

Scientists have long been able to change and/or modify single genes or small sets of genes. Most genetic alterations that people know about today are through engineering of crops, which involves adding or altering less than 10 genes out of the tens of thousands that are contained in most organisms or plants. Synthetic genomics is different in that scientists start with digital information in the computer, which allows for the design of entire synthetic chromosomes to replace existing chromosomes in cells. The first self-replicating synthetic bacterial cell constructed by scientists at the JCVI has more than 1 million base pairs of DNA, almost 1,000 genes, and involved the complete replacement of genetic material in the cell. More detail about the construction of this cell may be found below in the section "Creation of a Synthetic Bacteria Cell."

Why Construct a Synthetic Cell?

We believe that the ability to "write the genetic code", as we describe synthetic genomics, will enable a better understanding of the fundamentals of living cells. It will also enable us to direct cells and organisms to perform jobs, such as producing clean water or new biofuels that natural species cannot currently do to the needed scale and efficiencies.

Is This Research Creating a Synthetic Bacterial Cell "Creating Life from Scratch"?

No. We do not consider this to be "creating life from scratch"; rather, we are creating new life out of already existing life using synthetic DNA to reprogram the cells to form new cells with functions that are specified by the synthetic DNA.

What Are the Potential Applications of a Synthetic Cell? What Is the Impact of This Area of Science and the Resulting Technologies?

The work to create a synthetic cell will have a profound and positive impact on society in that it will enable a better understanding of the fundamentals of biology and of how life works. It will lead to new techniques and tools for advanced vaccine and pharmaceutical development, and will continue to enable the development of new biofuels and biochemicals. As well, these technologies could be used to produce clean water, new sources of food, textiles, human and veterinary drugs, bioremediation techniques, etc. More details on specific applications may be found below in the section "Beneficial Applications of Synthetic Genomics."

I believe, along with my teams at JCVI and the company Synthetic Genomics Inc (SGI), that this science has the potential to be a major wealth driver for societies. A recent report, "Synthetic Biology: Scope, Applications and Implications," from the Royal Academy of Engineering in the United Kingdom, states, "Synthetic biology has the potential to create another raft of major new industries, the development of which is likely to have profound implications for the future of the UK, European and world economies." http://www.raeng.org.uk/news/publications/list/reports/Synthetic biology.pdf

What Are the Risks Associated with Synthetic Organisms? Do the Risks of These Technologies Overshadow the Potential Benefits?

As with any new area of science, medicine or technology, synthetic genomics has the potential to be used for great societal benefit (biofuels, vaccines and pharmaceuticals, clean water, bioremediation, etc), but it could also be used for negative purposes.

So called dual use technologies need to be carefully discussed and reviewed both at the government level (Federal, state and local) both in the US and globally, as well as in accessible forums for bioethicists, educators, students, media and the public to learn about the science and understand these risks and benefits.

My teams at both the JCVI and at SGI have, as the leaders of this field, been driving these ethical and societal implications since the beginning of the research (for nearly 15 years).

The policy team at JCVI has completed study on options for governance of this field and is currently engaged in a study of the societal issues this work raises.

Many other countries are reviewing and discussing this area of science and as such numerous reports and reviews have also been conducted.

More detail may be found in the section below, "Ethical and Societal Implications/Policy Discussions about Synthetic Biology".

Does This Work Have Anything to Do with Humans/Human Research?

No. All synthetic genomics work to date, both at the JCVI and elsewhere, has focused on microorganisms.

It is anticipated that given how little is known about human biology that no applications of this work will or should be attempted in humans.

The way that this research will impact human lives is through the numerous applications such as new vaccines, pharmaceuticals, biofuels, etc.

What Safeguards/Controls Are in Place to Protect Against Accidental Environmental Release?

This is an extremely important question for this research and as such has been a major focus for the researchers at JCVI and SGI.

Building on the longstanding and successful history in molecular biology of millions of experiments engineering and using organisms such as *E. coli* to conduct research, JCVI and SGI researchers will be able to engineer synthetic bacterial cells so they cannot live outside of the lab or other production environments.

This is done by, for example, ensuring that these organisms have built in dependencies for certain nutrients without which they cannot survive. They can also be engineered with so called "suicide genes" that kick in to prevent the organism from living outside of the lab or environment in which they were grown.

Has There Been Any Review of This Work by the US Government, or by Any Other Organizations?

The synthetic genomics research at JCVI has undergone review at the highest levels of the US government. Beginning in 2003 with the publication of the research at JCVI in constructing the synthetic virus phiX174 ("Generating a synthetic genome by whole genome assembly: phiX174 bacteriophage from synthetic oligonucleotides." Smith et al, PNAS 2003 Dec 23;100(26):15440- 5. Epub 2003 Dec 2.), and including the most recent research and publication on creating the first self-replicating synthetic bacterial cell, the work has been reviewed by White House offices including the Office of Homeland Security and Office of Science and Technology Policy, the National Science Advisory Board for Biosecurity (NSABB), the Department of Energy, the National Institutes of Health, and others. As well, the work has been reviewed by independent bioethics groups since 1997. Senior US government officials including those at the NIH were briefed and allowed to review our study prior to publication.

What, If Any, Types of Legislation or Regulation Should Be Applied to This Area of Research?

We think that it is prudent, as is being proposed by the Department of Health and Human Services (HHS), to require DNA synthesis companies to screen synthesis requests against data on harmful agents. In 2004, JCVI's Policy team, along with the Center for Strategic & International Studies (CSIS) and the Massachusetts Institute of Technology (MIT) were funded by the Alfred P. Sloan Foundation to conduct a series of workshops and public sessions over a 20-month period to discuss the biosecurity and biosafety implications of synthetic genomics. Over the course of the study, the group explored the risks and benefits of the emerging technology, as well as possible safeguards to prevent abuse, such as bioterrorism. In October of 2007 the group published their findings in a report, outlining options for the field and its researchers moving forward.

More recently, in December of 2008, JCVI received funding, again from the Alfred P. Sloan Foundation, to examine ethical and societal concerns that are associated with the developing science of synthetic genomics. The ongoing research is intended to inform the scientific community as well as educate our policymakers and journalists so that they may engage in informed discussions on the topic.

What Are the Next Steps for this Research at JCVI?

The work to create the first self-replicating, synthetic bacterial cell was an important proof of concept. The team at JCVI has learned much from the nearly 15 years it has taken to get to this successful stage. From this proof of concept experiment the team is now ready to build more complex organisms with useful properties. For example, many researchers, including scientists at SGI, are already using available sequencing information to engineer cells that can produce energy, pharmaceuticals, and industrial compounds, and sequester carbon dioxide. The team at JCVI is already working on its ultimate objective, which has been to synthesize a minimal cell that has only the machinery necessary for independent life. Now that a cell can be synthesized from a synthetic genome, it now becomes possible for the team to test for the functionality of every essential gene in the genome. We can delete non-essential DNA regions from the synthetic genome and repeat transplantation experiments until no more genes can be disrupted and the genome is as small as possible. This minimal bacterial cell will enable a greater understanding of the function of every gene in a cell and a new vision of cells as understandable machines comprised of biological parts of known function.

Is This Research Patented?

Over the course of the 15 years it has taken to construct the first self-replicating synthetic bacterial cell, the team at JCVI has had to develop new tools and technologies to enable this research. SGI has funded the work at JCVI in exchange for exclusive intellectual property rights. SGI has filed 13 patent family applications on the unique inventions of the JCVI team. SGI believes that intellectual property is important for this kind of research and application development, as it is one of the best means to ensure that this important area of basic science research will be translated into key commercial products and services for the benefit of society. SGI intends to provide licenses to its synthetic genomics patents.

Creation of a Synthetic Bacterial Cell

The ability to sequence or "read" an organism's entire genome—the full repertoire of genes in that organism—has been possible for several decades and is now quite routine. Much can be learned about an organism by sequencing its genome. However, learning to write genetic code is crucial to truly understanding some of the most fundamental aspects of life. If scientists can write genetic code then it becomes possible to optimize certain functions in organisms that would be beneficial for society. With these ideas in mind, we set out to create a synthetic bacterial cell. The work has its roots in 1995 and 1999 publications on *Mycoplasma genitalium*, but the quest to develop the first synthetic bacterial cell began in earnest in 2003.

May 21, 2010 Science Publication

On May 21, 2010, the JCVI synthetic genomics team of nearly 25 researchers, led by me, Hamilton Smith, Clyde Hutchison, John Glass, and Dan Gibson, published results detailing the first cell constructed in the lab using only synthetic DNA. The work was published online in the journal *Science* and details the work to chemically synthesize the 1.08 million base pair genome of the bacterium *Mycoplasma mycoides*.

This and previous breakthrough work by JCVI researchers was funded by Synthetic Genomics Inc. The US Department of Energy also funded early work in this area, particularly the work to construct the synthetic phiX174 published in 2003.

Using previously published techniques and breakthroughs with the genetic system of yeast and of genome transplantation, the team put chemically synthesized pieces of the *M. mycoides* DNA into yeast which assembled the bacteria's genome. Then, the *M. mycoides* genome was transplanted into *Mycoplasma capricolum* and "booted up" to create a new synthetic version of *M. mycoides*.

Steps involved in building the synthetic *M. mycoides* are as follows:

1) First, the JCVI team designed 1,078 specific cassettes of DNA that were 1,080 base pairs long, with overlaps of 80 base pairs (bp) at their ends to aid in building the longer stretches of DNA. These were made according to JCVI's specifications by the DNA synthesis company, Blue Heron Biotechnology.
2) Then the team employed a three stage process using yeast to build the genome from these 1,078 cassettes. The first stage involves taking 10 cassettes of DNA at a time to build 10,000 bp long segments. In the second stage, these 10,000 bp segments are taken 10 at a time to produce eleven 100,000 bp long segments. Finally, all 11 segments are assembled into a complete synthetic genome as an extra chromosome in a yeast cell, by using yeast genetic systems.
3) The complete synthetic *M. mycoides* genome is then released from the yeast cell and transplanted into *M. capricolum* recipient cells that have had the gene for a restriction enzyme removed. Following incubation, viable *M. mycoides* cells are produced in which the only DNA present is the synthetic genome. These cells are controlled only by that synthetic genome.

Scientific Milestones on the Quest to Create the First Synthetic Bacterial Cell

1995: After sequencing the M. genitalium genome (published in 1995), my colleagues and I began work on the minimal genome project. This area of research, trying to understand the minimal genetic components necessary to sustain life, started with M. genitalium because it is the bacterium with the smallest genome known that can be grown in pure culture. This work was published in the journal Science in 1999.

2003: Drs. Venter, Smith, and Hutchison (along with JCVI's Cynthia Andrews-Pfannkoch) made the first significant strides in the development of a synthetic genome by assembling the 5,386 base pair genome of bacteriophage phiX 174. They did so using short, single strands of synthetically produced, commercially available DNA (known as oligonucleotides) and using an adaptation of polymerase chain reaction (PCR), known as polymerase cycle assembly (PCA), to build the phiX genome. The team developed methods that allowed the synthetic phiX to be produced in just 14 days. This work was published in the Proceedings of the National Academy of Sciences (PNAS).

2007: JCVI researchers led by Carole Lartigue, Ph.D., announced the results of work published in the journal Science, which outlined the methods and techniques used to change one bacterial species, M. capricolum, into another, M. mycoides , by replacing one organism's genome with the other's genome. Genome transplantation was the first essential enabling step in the field of synthetic genomics as it is a key mechanism by which chemically synthesized chromosomes can be activated into viable living cells.

January 2008: The second successful step in the JCVI team's effort to create a cell controlled by synthetic DNA was completed when Gibson et al. published in the journal Science, the synthetic M. genitalium genome.

December 2008: Gibson et al. published a paper in Proceedings of the National Academy of Sciences (PNAS) describing a significant advance in genome assembly in which the team was able to assemble in yeast the whole bacterial genome, M. genitalium, in one step from 25 fragments of DNA. The work was funded by the company Synthetic Genomics Inc. (SGI). At this point the team is still working to boot up the M. genitalium synthetic cell using all the knowledge gleaned from their previous work.

2009: JCVI researchers published results describing new methods in which the entire bacterial genome from M. mycoides was cloned in a yeast cell by adding yeast centromeric plasmid sequence to the bacterial chromosome and altered in yeast using yeast genetic systems. This altered bacterial chromosome was then isolated from yeast and transplanted into a related species of bacteria, M. capricolum, to create a new type of M. mycoides cell. This was the first time that genomes were transferred between branches of life—from a prokaryote to eukaryote and back to a prokaryote. The research was published by Lartigue et al. in Science.

Synthetic Genomics and Synthetic Biology Defined

Synthetic genomics is a new capability that engages in the design and assembly of genes, chromosomes, and potentially entire multi-chromosome genomes. The basic units of construction are chemically synthesized oligonucleotides (called oligos). Oligos are short strings of DNA formed from the four nucleotide bases (i.e., A, C, G, and T).

Although the terms are sometimes used interchangeably, synthetic genomics differs from the more widely known synthetic biology in the scale of changes that can be made and in the kinds of experiments that it enables. Synthetic biology, by its community's view, is derived from engineering principles and is focused on the design and construction of biological parts (genes, pathways), devices (multiple parts), and systems (multiple devices). The chief aim of synthetic biology is to provide standardized sets of 'parts' that can be joined together in new ways in a living organism. Synthetic genomics technologies, on the other hand, provide the capability to build whole genomes and can examine how best to organize them.

While methods and tools for conducting synthetic biology have been available for many years, synthetic genomics is a completely new capability developed at JCVI.

Our program in synthetic genomics has developed a set of techniques that are fundamental to engineering an organism in its entirety. Two key features of our synthetic genomics capabilities are: 1) rapid assembly of DNA molecules up to millions of base pairs in size (a million base pairs can code for ~1000 genes) and 2) "combinatorial" reconstruction of genomes, that is, novel genetic arrangements can be produced and assayed quickly. The application of rational engineering principles to the construction of combinatorial libraries (collections of pieces of DNA put together in different arrangements)—followed by high-throughput screens to select for optimal arrangements— ensures hundreds-to hundreds of thousands of competing designs can be examined in parallel.

One of the major advantages of synthetic genomics over classical biotechnology techniques— such as recombinant DNA—is that there is no need to have access to a physical supply of a particular DNA sequence. Sequence fragments are simply created de novo by chemical synthesis and assembled into entire chromosomes and organisms.

This ability to synthesize (write) DNA and use it in the construction of new cells can catalyze a major change in what organisms can be engineered to do. Importantly, it will also increase our understanding of microbial life processes.

Not only can new cells types be created but existing natural systems can be exhaustively probed to reveal the inner workings and properties displayed by living organisms.

BENEFICIAL APPLICATIONS OF SYNTHETIC GENOMICS

Synthetic genomics will make a unique or significant contribution as an enabling technology that is changing the nature of basic biological research; and as a powerful tool of applied biotechnology with the potential for developing new or improved applications for human health (including new pharmaceuticals and faster development of vaccines), biological sources of liquid transportation fuels, the manufacturing of other bio-based products, and environmental surveillance.

Synthetic genomics is today changing the nature of *basic molecular biological research.* As an enabling technology, DNA synthesis has already proved to be a significant time saver by shortening the time needed for experiments compared to time-consuming recombinant DNA techniques. As DNA synthesis becomes ever less expensive, researchers will be able to use synthetic genomics to rapidly change the DNA sequence of various genes or whole genomes, allowing them to understand basic cellular functions in a rigorous way. For example, various laboratories are beginning to use synthetic genomics (specifically, the combinatorial reconstruction of genomes) to understand the mechanisms of evolution at the molecular level, to define regulators of specific genes or gene pathways and to establish, at the molecular level, the minimal requirements for life. Without synthetic genomics, investigators can only manipulate one or at most a few genes in any given experiment, resulting in a relatively slow discovery process.

These laboratory techniques can also be applied to beneficial products. Drugs, vaccines, and modified microbes for use in humans are all important targets of applied research using synthetic genomics. The capability to make subtle changes at the DNA sequence level may lead to more efficient research and production of *vaccines for human and animal health* and related *diagnostics*. Currently, scientists are working on ways to use synthetic genomics technologies for the mitigation of influenza epidemics with the eventual ability to generate vaccines more rapidly than they are currently being generated. These technologies could be applied to several steps in the vaccine development process, resulting in moderate to significant time savings compared to current methodologies. Additionally, the ability to assemble and mutate sequences rapidly could allow for the development of broadly protective vaccines against viruses that themselves are diverse and variable, such as the viral causative agents of severe acute respiratory syndrome (SARS) and hepatitis C.

The JCVI has recently been funded by the NIH to use our new synthetic DNA tools to build synthetic segments of every known flu virus so that we can rapidly build new vaccine candidates in less than 24 hours. We are also being funded to see if we can take sets of genes out of bacteria to design new antibiotic synthetic pathways to make chemical compounds that are currently too complex for chemists to make. With the extensive research already underway in this new field of synthetic biology, there will be thousands of new developments that we cannot imagine today. DNA synthesis techniques have already been applied in research on new or improved drugs. For example, the antimalarial drug artemisinin is naturally produced in the plant Artemisia annua through a complex metabolic pathway that cannot feasibly be reconstructed in yeast using conventional biotechnological methods. Scientists have been able to synthesize an artemisinin precursor (which is then subject to chemical modification to make the final product) and are in the process of learning how to scale up this production to make the drug widely and relatively inexpensively available. This type of modification is likely to be applicable to a wide variety of drugs. Synthetic genomics could also contribute to the search for *carbon-neutral energy sources*. A major application of synthetic genomics could be in overcoming biological barriers to cost-effective production of biofuels. There are several major initiatives in alternative or substitute fuel production research. One promising approach now is to engineer photosynthetic algae (either microalgae or blue-green algae) that are already relatively efficient at converting carbon dioxide into oils so that they carry out this process at a scale that is commercially viable. While biofuels from algae may be the best current target for alternative fuels, *consolidated bioprocessing* (CBP) of cellulosic biomass to ethanol is a possible route as well, and may be preferable in some settings. Scientists are trying to engineer a single organism to include all the multiple steps needed to produce ethanol from cellulose. While the use of synthetic genomics to produce all of the enzymes needed for CBP is not the only technique available, it is among the most promising. If successful, CBP might be able to produce ethanol at a cost competitive with gasoline. Sometimes called "white biotechnology," *biobased manufacturing* is becoming a reality. Plants and microbes are being engineered to produce raw materials that can be used to manufacture products that today are typically petroleum based. The expectation is that biologically based manufacturing will lead to more environmentally friendly products and methods of production. For example, the environmental impacts of plastic manufacturing might be lessened through the judicious use of bioengineering of metabolic pathways using synthetic genomics as one tool.

Finally, synthetic genomics could be applied to constructing microbes or other organisms that would act as *detectors* of toxins, chemicals, or even other (pathogenic) microbes in routine or bioterrorism surveillance. This could aid international health organizations greatly in early detection of emerging diseases.

Ethical and Societal Implications/Policy Discussions About Synthetic Biology

At JCVI, we consider the ethical and societal implications of the work to be as important as the scientific research. We examined ethical concerns before beginning any actual experiments or research into constructing a minimal genome or the work to construct the first synthetic cell. Here is an outline of the important work that JCVI has undertaken since 1995.

1995-1999: *Mycoplasma genitalium* and the Minimal Genome Project

Research on the minimal genome started in 1995 after the publication of the *Mycoplasma genitalium* genome at the legacy JCVI organization, The Institute for Genomic Research (TIGR).

This organism has the smallest genome of a self-replicating organism, prompting my team and me to wonder if *M. genitalium* could be a platform to determine the minimal set of genes that could still sustain cellular life.

This notion and the research plan to test it underwent a thorough ethical review by a panel of experts at the University of Pennsylvania (Cho et al., Science December 1999:Vol. 286. no. 5447, pp. 2087 – 2090).

The panel's independent deliberations, published in Science along with the scientific minimal genome research, concluded that there were no strong ethical reasons that should prevent the team from continuing research in this field as long as they continued to engage in public discussions.

JCVI Work on phiX174 Synthesis: The First Synthesis of a Non-Pathogenic Virus

In 2003, before publishing the results in PNAS ("Generating a Synthetic Genome by Whole Genome Assembly: phi X174 Bacteriophage from Synthetic Oligonucleotides"), our team of scientists from JCVI contacted several Government agencies, including the US Department of Energy (DOE), the White House Office of Science and Technology Policy (OSTP), and the National Institutes of Health, to discuss any potential repercussions of the findings. After a series of meetings (which also included Department of Homeland Security representatives) discussing the method presented in the paper, the findings were released at a press conference hosted by DOE in conjunction with the Secretary of Energy, Spencer Abraham.

JCVI Policy Team

Shortly after, in 2004, JCVI's Policy team along with the Center for Strategic & International Studies (CSIS) and the Massachusetts Institute of Technology (MIT) were funded by the Alfred P. Sloan Foundation to conduct a series of workshops and an invitational public session over a 20-month period to discuss the biosecurity and biosafety implications of synthetic genomics.

Over the course of the study, the group explored the risks and benefits of the emerging technology, as well as possible safeguards to prevent abuse, such as bioterrorism. In October of 2007 the group published their findings in a report, outlining options for the field, its researchers, science administrators, and policymakers.

More recently in December of 2008, JCVI (in collaboration with social science researchers from Michigan State University and the Alberta Health Law Institute) received funding from the Alfred P. Sloan Foundation to examine ethical and societal concerns that are associated with the developing science of synthetic genomics.

The ongoing research is intended to inform the scientific community as well as educate policymakers and journalists so that they may engage in informed discussions on the topic.

Ongoing Activities: Lectures, Media, Briefings for Congress and Executive Branch Officials

The JCVI team and I routinely give public lectures and presentations around the globe to both scientific and lay audiences, members of congress, schools, and other organizations. The team and I also conduct many interviews with global media (online, print, video, radio, etc.) about our work and the implications and applications.

Over the last three years the team has made several trips to Capitol Hill to brief more than 50 members of Congress. The most recent work published on the first synthetic bacterial cell published in *Science* has been reviewed by OSTP, Department of Homeland Security, the NSABB, etc. The team supports and has asked for continued review and discussion about their research.

SELECTED STUDIES OF THE SOCIETAL, ETHICAL, AND POLICY CONCERNS

To help the Committee in your deliberations, I have assembled a list of key studies of the societal, ethical, and policy concerns associated with synthetic genomics and synthetic biology.

Completed Studies and Reports from the United States

- Cho MK, Magnus D, Caplan AL, McGee D, and the Ethics of Genomics Group. 1999. Ethical Considerations in Synthesizing a Minimal Genome. Science 286: 2087-2090. http://www. sciencemag.org/cgi/content/short/286/5447/2087. This was the earliest study of the societal and ethical implications of synthetic genomics. Funded by an unrestricted grant from The Institute for Genomic Research Foundation (TIGR), a legacy organization of today's JCVI. The study was performed in parallel with research to define a minimal bacterial genome.
- United States Department of Energy, Office of Science, Biological and Environmental Research Advisory Committee, 2004. Synthetic Genomes: Technologies and Impact. http://www.science. doe.gov /ober/berac/synbio.pdf. Report of a DOE advisory group on the potential benefits and concerns associated with synthetic genomic technologies.
- National Science Advisory Board for Biosecurity (NSABB) 2006. Addressing Biosecurity Concerns Related to the Synthesis of Select Agents. http://oba. od.nih.gov/biosecurity/pdf/Final_NSABB_Report_on_Synthetic_Geno mics.pdf The Subcommittee on Synthetic Genomics of the NSABB prepared this report on security

issues related to the construction of select agents using synthetic genomics technologies.

- Garfinkel MS, Endy D, Epstein GL, and Friedman RM, Synthetic Genomics: Options for Governance, 2007. http://www.jcvi.org /cms /fileadmin/site/research /projects/syntheticgenomics-report/synthetic-genomics-report.pdf. Report focuses on the biosecurity and biosafety concerns associated with synthetic genomics and presents and evaluates 17 policy options for consideration by policymakers. The two-year study, funded by the Alfred P. Sloan Foundation, was prepared by Michele Garfinkel and Robert Friedman, JCVI; Drew Endy, MIT; and Gerald Epstein, Center for Strategic & International Studies.
- National Academies/OECD/Royal Society, 2009. Opportunities and Challenges in the Emerging Field of Synthetic Biology: A Symposium http://sites. nationalacademies.org/PGA/stl/PGA_050738. This two-day symposium, funded by the Sloan Foundation, NSF, and BIO brought together biologists, social scientists, and policy experts to educate each other and to explore possibilities for trans-Atlantic collaborations.

Ongoing US-based studies

- Synthetic Genomics: Scientists' Understanding of Society's Concerns, Society's Understanding of the Science and Scientists http://www. sloan.org/assets /files/press/alfred p sloan foundation funds new synthetic biology initiative to examine societal issues.pdf JCVI's current study on the societal implications of synthetic genomics, funded by the Alfred P. Sloan Foundation (2009-2010). Garfinkel and Friedman from JCVI, in conjunction with Lori P. Knowles, University of Alberta, Health Law Institute and Paul B. Thompson, Michigan State University, Department of Philosophy, are examining the sometimes differing views of society and scientists with respect to synthetic genomics. Also examines regulatory issues in the US and the EU.
- Synthetic Biology Project: Ensuring Benefits are realized through Responsible Development http://www.synbioproject.org/ The Woodrow Wilson International Center for Scholars established this project as an initiative of the Foresight and Governance Program with a grant from the Alfred P. Sloan Foundation. The project aims to identify gaps in knowledge about risks, to understand public perceptions about the field, and to explore governance options to promote innovation while ensuring safety.
- Ethical Issues in Synthetic Biology: Non-Physical Moral Harms and Public Policy http://www.thehastingscenter.org/Research/Detail.aspx? id=1548 Funded by the Alfred P. Sloan Foundation, this project aims to identify non-physical concerns about and potential consequences of synthetic biology, including how to incorporate these concerns into public policy discussions.
- Synthetic Biology Engineering Research Center (SynBERC) http://www.synberc.org /content/articles/human-practices SynBERC is a multi-institutional research group

funded by National Science Foundation to explore a number of engineering issues in synthetic biology.

US Government actions

- 2009. Federal Register Notice: Department of Health and Human Services, National Institutes of Health, Office of Biotechnology Activities. Recombinant DNA Research: Proposed Actions Under the NIH Guidelines for Research Involving Recombinant DNA Molecules (NIH Guidelines). http://oba.od.nih.gov/oba/rac /ProposeRevisions NIHGuidelines-March-4-2009.pdf Considers whether synthetic DNA is identical to recombinant DNA with respect to NIH Guidelines and thus whether language in the Guidelines needs to be changed. Public comments are currently under review.
- 2009. Federal Register Notice: Department of Health and Human Services, Office of the Secretary. Screening Framework Guidance for Synthetic Double-Stranded DNA Providers. http://www.gpo.gov /fdsys/pkg/FR-2009-11-27/pdf/E9-28328.pdf. Provides guidance to firms that supply synthetic DNA with respect to screening orders and customers for malicious intent. Public comments are currently under review.

Completed studies and reports from the United Kingdom and Europe

- De Vriend H, for the Rathenau Institute. 2006. Constructing Life: Early social reflections on the emerging field of synthetic biology. http://www.cisynbio.com/pdf /Constructing_Life_2006.pdf. Early and rigorous description of the constellation of societal issues that may be raised by synthetic biology.
- International Association Synthetic Biology Code, 2009. The IASB Code of Conduct for Best Practices in Gene Synthesis. http://www.ia-sb.eu/tasks/sites/synthetic-biology/assets/File/pdf/iasb code of conduct final.pdf. A suggested code of conduct for DNA synthesis firms, drafted by members of the IASB consortium. IASB is European-based; the process to draft this Code of Conduct included US firms.
- Synbiosafe (European Commission 6th Framework Program, Project on Synthetic Biology Safety and Ethical Aspects). http://www.synbiosafe.eu/. Three major products, all edited/directed by M. Schmidt, Synbiosafe manager: a book (Synthetic Biology: The Technoscience and Its Societal Consequences), a documentary film (SYNBIOSAFE: Safety and Ethical Aspects of Synthetic Biology), and a special issue of Systems and Synthetic Biology (Societal Aspects of Synthetic Biology)
- UK Parliamentary Office of Science and Technology, 2008. POSTnote http://www. parliament.uk/documents/upload/postpn298.pdf. This document discusses possible applications and risks of new synthetic biology, including policy options for governance and development of the field.

Ongoing Studies in Europe, 7th Framework Program

- SYBHEL (Synthetic Biology for Human Health: Ethical and Legal Issues) http://sybhel.org/. This is one of just a few ethics and policy projects worldwide to focus solely on the impacts of synthetic biology technologies with respect to human health.
- Synth-Ethics (Ethical and Regulatory Issues Raised by Synthetic Biology) http://www.synthethics.eu/. This is a general project focused on safety, security, and notions of life, looking both at Europe generally and within specific countries.

Synthetic Biology Periodic Meetings: SB 1.0, 2.0, 3.0, 4.0, x.0....

The synthetic biology community holds recurring international meetings that include ethicists and social scientists with general interests in the research. Each meeting has dedicated time to presentations on societal impacts and issues.

- SB 1.0, 2005. http://syntheticbiology.org/Synthetic_Biology_1.0.html
- SB 2.0, 2006. http://webcast.berkeley.edu/event_details.php?web castid=15766
- SB 3.0, 2007. http://www.syntheticbiology3.ethz.ch/monday.htm
- SB 4.0, 2008. http://sb4.biobricks.org/agenda/sb4_agenda.pdf

I thank you for the opportunity to testify before you today. I welcome any questions that you may have.

In: Synthetic Biology
Editors: Roger E. Hamilton and Laura M. Morales

ISBN: 978-1-62257-207-6

Chapter 4

Testimony of Dr. Jay Keasling, University of California at Berkeley. Hearing on "Effects of Developments in Synthetic Genomics"*

Chairman Waxman, Ranking Member Barton, Chairman Emeritus Dingell and distinguished members of the Committee, thank you for inviting me to testify at this important hearing. Research and scientific innovation is the key to America's long-term health and economic vitality. My scientific colleagues and I appreciate your shining a light on this new and exciting field of research and its great potential to benefit the world.

My name is Jay Keasling. I am a Professor in the Departments of Chemical Engineering and Bioengineering at the University of California at Berkeley; Acting Deputy Director of Lawrence Berkeley National Laboratory, a U.S. Department of Energy (DOE) multipurpose national laboratory operated by the University of California; Chief Executive Officer of the DOE-funded Joint BioEnergy Institute; and Director of the Synthetic Biology Engineering Research Center.

Synthetic biology is the engineering of biology to solve important problems. It is basic science with a focus on application. This emerging field of fundamental science has great potential for developing solutions to large-scale societal challenges.

Although most people are familiar with "genetic engineering" or "molecular biology," synthetic biology uses an approach and tools that differ significantly from both. The differences are in the approach and the tools utilized. For example, early molecular biologists cobbled together natural biological components and hoped that the engineered system would work. Assembling the components was challenging, and as a result, engineered organisms rarely functioned as desired. Today, synthetic biologists have improved the reliability and safety of engineered organisms by assembling standardized well-characterized components from existing well-studied organisms much like how one might assemble a computer from standard components such as a hard drive, sound card, motherboard, and power supply.

My research focuses on engineering microorganisms to produce pharmaceuticals and biofuels. In my lab, we use well-known microorganisms such as *E. coli* and yeast that

have been widely used for many years by the biotechnology industry. We introduce into these microorganisms DNA that encodes biological components such as metabolic pathways

---

* This is an edited, reformatted and augmented version of testimony given on May 27, 2010 before the House Committee on Energy and Commerce.

that enable the organism to transform inexpensive sugars into valuable, useful products. These products include drugs for diseases that afflict people in the developing world and carbon-neutral biofuels to reduce our dependence on foreign oil.

When I started my career in 1992 as an assistant professor at the University of California at Berkeley, biological components were relatively crude, making the engineering of microorganisms time-consuming and costly. But my colleagues and I had the idea that one could engineer microorganisms into chemical factories that produce nearly any important chemical from sugar. Unfortunately, there were few tools available to us at the time. So we began by developing tools to accurately produce a chemical of interest by controlling the expression of genes that had been transferred into cells. At that time, there was no name for what we were doing, but now it is known as synthetic biology.

Thanks to the National Science Foundation's investments in the Synthetic Biology Engineering Research Center (SynBERC), my colleagues and I are now establishing standards for the engineering of biology and creating and characterizing biological components that can be readily assembled to solve important problems. SynBERC brings together many of the pioneers (biologists and engineers from world-class institutions) of synthetic biology who are laying the foundation for the new field of synthetic biology. We are working together to construct standard, reliable, and safe building blocks that can be used in a myriad of applications. We are also studying safety, security, preparedness, and ethics issues around these powerful technologies to ensure they are used safely and wisely.

One of the most important and well-known applications of synthetic biology has been our work on engineering microorganisms to produce the anti-malarial drug artemisinin. There are 300 to 500 million cases of malaria at any one time. One to three million people die from the disease each year, and 90 percent of those are children under the age of five. While conventional quinine-based drugs are no longer effective, plant-derived artemisinin combination therapies are highly effective but cost prohibitive for much of the world.

To decrease the cost of artemisinin, we engineered a microorganism to produce a precursor chemical to the drug by transferring the genes responsible for making the drug from the plant to the microorganism. Through funding from the Bill and Melinda Gates Foundation, we completed the science in three years, largely due to ready access to well-characterized biological components. The microbial production process has been licensed by Sanofi-Aventis, which will scale the process and produce the drug within the next two years; selling it at cost in the developing world.

The process for producing artemisinin is akin to brewing beer. The microorganism takes in a sugar and secretes a precursor to artemisinin rather than alcohol, which the yeast would produce naturally from sugar. We predict that when this process is fully implemented, the drug produced by this engineered organism could save a large fraction of the two million or so children who die every year from malaria. Launching this process in 2011 or 2012 is crucial, as the plant-derived version of the drug will soon be in short supply.

Because the engineering of biology is time-consuming and unpredictable, the artemisinin project required $25 million in funding and roughly 50 people working on the project for the past three years. Through synthetic biology, we hope to make the engineering of biology more predictable and reliable, thereby reducing the cost to develop medicines and other useful products ranging from chemicals and fuels to consumer and commercial products.

Speaking of fuels, ethanol, which has been widely used as an oxygenate in gasoline and is the majority component of E-85, is not an ideal gasoline replacement. A gallon of ethanol packs only two-thirds of the energy of a gallon of gasoline. Ethanol is corrosive to engines and pipelines and requires energy-intensive distillation to purify. As such, its use would require significant changes to our transportation infrastructure, including replacing pipelines and automobiles at a significant cost.

Fortuitously, artemisinin is a hydrocarbon, a fundamental building block for fuel. We are now re-engineering the artemisinin-producing microbes to produce drop-in biofuels. That is, through advances in synthetic biology, we can engineer these same safe, reliable, industrial microorganisms to produce biofuels that will work within our existing transportation infrastructure.

The Joint BioEnergy Institute (JBEI), a Lawrence Berkeley National Laboratory–led scientific partnership between Sandia National Laboratories, the University of California campuses at Berkeley and Davis, the Carnegie Institution for Science, and the Lawrence Livermore National Laboratory, is exploring the potential of synthetic biology to advance the development of the next generation of biofuels—liquid fuels derived from the solar energy stored in plant biomass. JBEI is one of three DOE Bioenergy Research Centers funded by the Office of Biological and Environmental Research with the Department's Office of Science.

The approach of JBEI is to use the advances in synthetic biology to engineer microorganisms to transform sugars derived from cellulosic biomass and starch into hydrocarbon-based biofuels that have the same qualities as the fuels that are currently derived from petroleum. These new, advanced biofuels reduce the production of greenhouse gases, as they are derived from plants that use sunlight and atmospheric carbon dioxide to grow. These biofuels will reduce our dependence on foreign oil and could rejuvenate the U.S. agriculture economy, potentially making the American Midwest the new Middle East. My research is the foundation for two California-based advanced biofuel companies that are currently employing hundreds of people.

JBEI researchers have used synthetic biology and metabolic engineering techniques in *E. coli* and yeast to produce these advanced "drop-in" fuels that perform better than ethanol. The scientists are redirecting central metabolic, fatty acid, and cholesterol biosynthetic pathways to produce candidate gasoline, diesel, and jet fuel molecules. In work performed collaboratively with a Bay Area–based advanced biofuel company, my laboratory recently reported the engineering of *E. coli* to produce a biodiesel from the sugar polymer hemicellulose, a major component of plant biomass. The engineered microorganism secreted enzymes that digested hemicellulose, imported the sugar, transformed the sugar into diesel fuel, and secreted the diesel into the fermentation broth. The diesel floats to the top of the tank where the engineered *E. coli* are grown and can be skimmed off and used with very little purification. This engineering feat would not have been possible just a few years ago and certainly not without the recent advances in synthetic biology.

JBEI also has developed a new metabolic pathway that potentially could produce both advanced fuels and other molecules that might otherwise be produced from petroleum, paving the way to replace a significant portion of petroleum-based products with sugar-based products.

Very similar technologies are being used at JBEI to engineer plants to become efficient producers of cellulose with minimal input of water and fertilizer. Indeed, the advances in synthetic biology will allow us to have plentiful food to feed the population of the U.S. and the world as well as biomass for biofuels.

JBEI is also looking at the development of new and better enzymes. To break down the rugged lignocelluloses of biomass material, JBEI researchers have analyzed microbial communities in Puerto Rican rainforest soils that boast some of the planet's highest rates of biomass degradation. To perform the analysis, scientists used the Phylochip, a credit card–sized microarray developed at Lawrence Berkeley National Laboratory that can quickly detect the presence of up to 9,000 microbial species in samples. Using bags of switchgrass as "microbe traps," the researchers conducted a census of these soil microbes to identify the most efficient biomass-degrading bacteria and fungi. Understanding how these microbes work may provide synthetic biology solutions to more efficient and affordable deconstruction of biomass for advanced biofuels production.

Many other applications could benefit from advances in synthetic biology, including nitrogen-fixing crops that do not need ammonia-based fertilizers, microorganisms engineered to produce all of the chemicals currently produced from petroleum, and entirely new classes of drugs to fight cancer, infections of multidrug-resistant bacteria such as those that cause tuberculosis, and a host of other diseases.

I hope that my testimony has illustrated for you the remarkable potential of synthetic biology and the important role that it has to play in our nation's research and innovation enterprise. Your actions and the support of Congress will determine whether the efforts described today are ultimately successful. This is a marathon, not a sprint, and requires consistent and continuous nourishing and care.

We are very encouraged by the language adopted by the House Committee on Science and Technology regarding synthetic biology in the America COMPETES Act, and stand ready to assist Congress in any way we can as you explore and learn more about this exciting research area.

Finally, thank you, again, for holding this important hearing and for inviting me to participate. Please let me know if I may ever be of any assistance. I will be happy to answer any questions.

In: Synthetic Biology
Editors: Roger E. Hamilton and Laura M. Morales
ISBN: 978-1-62257-207-6

Chapter 5

TESTIMONY OF DR. DREW ENDY, STANFORD UNIVERSITY. HEARING ON "EFFECTS OF DEVELOPMENTS IN SYNTHETIC GENOMICS"*

Good Morning, Chairman Waxman, Ranking Member Barton, and Members of the Committee.

My name is Drew Endy. I am an Assistant Professor in the Department of Bioengineering at Stanford University, President of the BioBricks Foundation, and Director of the BIOFAB: International Open Facility Advancing Biotechnology (BIOFAB). I serve on the Committee on Science, Technology and Law at the National Academies and am a recent nominee to the National Science Advisory Board for Biosecurity. My own work and that of my students is the direct result of sustained public funding for basic science and engineering research from the NIH, NSF, and DOD, and for which we are grateful.

Synthetic biology has been called "extreme genetic engineering" by civil society organizations. This label is true but only in relation to the past 35 years of biotechnology that follow the invention of recombinant DNA technology and other early tools.

Speaking as an engineer, the facts today suggest that we are extremely bad at engineering biology.

For example, in my own lab we are working to understand and engineer how cells make decisions, store information, and communicate. One current "holy grail" is to implement a genetically encoded 8-bit information storage system. Our deliverable is similar to a computer's memory chip or a USB flash drive that you might use with a digital camera, except for two major differences. First, our system will only store 8 bits, which is 8 billion times less than what you could store on an electronic memory stick available today from Walmart for $20. Second, our system is made from proteins and DNA that function inside living cells. The system works by controlling enzymes that flip DNA back and forth; a stretch of DNA pointing "left" means "0" while "right" means "1". We will use our system to study and control cancer, aging, and development. For example, we plan to create combinatorial counters that track the number of times cells divide, and explore the possibility of building into cells additional "fail-safes" that prevent out-ofcontrol replication, such as during cancer. Practically, the design of our first 8-bit combinatorial counter requires that we combine the

* This is an edited, reformatted and augmented version of testimony given on May 27, 2010 before the House Committee on Energy and Commerce.

DNA sequences for at least 48 genes encoding the various DNA flipping enzymes with as many more control elements. In total we need to design, build, and test about 100,000 base pairs, or letters, of highly engineered DNA. Using the best tools available it has taken us over one year to get the molecular pieces that comprise our first bit working.

The high cost and uncertainty of doing genetic engineering research has big impacts. For example, ~99% of all engineered genetic "programs" today are encoded by less than 20,000 base pairs of designer DNA. As a second example, the NIH is thought to spend ~5% of its total annual budget supporting researchers who then spend up to 50% of their time manipulating DNA by hand. Thus, while most of the attention is focused on the applications or ethics of biology and biotechnology, it is also important to look at the tools, processes, and human practices that comprise the work itself. This is where synthetic biology has a powerful role to play.

For example, we have heard today how researchers at the J. Craig Venter Institute (JCVI) reconstructed a 1 million base genome using DNA synthesis. As a related example, in 2005, researchers in Japan constructed a composite genome approaching 8 million base pairs in length, starting from natural DNA fragments. The scale of these genome construction projects is ~10 to 100-fold beyond what's required by most research projects. Very simply, today, if every publicly funded biomedical or biotechnology research team had direct access to a gene or genome printer, most researchers could focus their full attention on the challenges of understanding and applying biology to solve problems instead of spending the majority of our time "bashing" DNA.

As a second example of synthetic biology in practice, just because we have DNA printers does not mean that we will have much useful to say. We need to also discover or invent the languages and grammars that enable us to write more powerful genetic programs, moving from today's simple declarative statements – "synthesize lots of insulin" – to tomorrow's short stories and novels – "identify, attack, and destroy the tumor in this patient, and then differentiate and re-grow into healthy replacement tissue." Here is where old but powerful engineering ideas based on standardization and abstraction are starting to have an impact. For example, the public-benefit BIOFAB facility in Emeryville CA has a two-year goal of producing a first "operating system" supporting large-scale and reliable genetic programming in the bacterium *E. coli*, which is a well-studied model organism and "workhorse" of industrial biotechnology. We estimate our first cellular operating system will include ~3,000 standard biological parts encoding different cellular control functions. We intend to make this cellular operating system freely available so that all researchers can more quickly and reliably engineer useful genetic programs.

At this point, let me acknowledge that one characteristic of synthetic biology is how quickly some of the core tools continue to change. For example, over the past five years, the length of the longest genome synthesized from scratch has increased by a factor of ~100. Thus, five years from now, we might expect that further advances in synthetic biology will enable the construction of ~100 million base pair genomes, a length nearly sufficient to encode worms and flies. To be clear, the capacity to construct genomes at such scales will not mean that we know how to "weave a worm" or "fly a fly." Rather, it guarantees that we will remain challenged to become orders of magnitude better at the basic science and engineering of biology for the foreseeable future. We will also be challenged to keep pace with developments and to sustain constructive dialog and policy-making across a diversity of concerns, values, and perspectives.

Regarding the impact of synthetic biology on national energy policy generally let me make two points. First, in very rough terms, life on Earth is thought to handle 100 terawatts of energy; human civilization uses 20 terawatts. Although it might appear that biology presents us with a 5-fold surplus as a potential energy source, we depend on the energy flowing through biology to provide for many other obviously essential needs, from ecosystems to ourselves. Thus, future large-scale deployments of synthetic biology-based technologies will need to be proactively coupled to the constructive resolution of matters involving resource utilization and land use politics.

Second, from an energy perspective, the ultimate value of biology as a manufacturing platform that addresses our nation's energy needs goes beyond the production of bulk commodity products such as liquid fuels. For example, by the early 1980s, the enzymes used in laundry detergents to treat stains had been engineered to work at cold-water wash temperatures by companies such as Genencor, Inc., resulting in the potential reduction of hot water heating bills amounting to 100,000 barrels of oil per day, nationwide. Stated differently, the "energy impact" of a single engineered protein integrated upstream into our daily lives via a laundry detergent is greater than the current oil spill in the Gulf of Mexico, and is roughly equivalent to the volume of biofuel that could be produced using 1/2000th of our crop land. Synthetic biology brings us many more opportunities to better partner with biology in reducing our energy needs and net impact on the natural environment.

In closing, let me return to the work that has brought us here and briefly sketch some of its significance from a policy and governance perspective. What changes now that it is possible to construct a replicating cell from a synthesized genome?

From a safety perspective, we inherit a tradition of practical success from the genetic engineering generation. Via synthetic biology many more people will seek to work with and use biotechnology. For example, thousands of young engineering students now labor to design and synthesize simple DNA programs via the iGEM competition. We must renew and advance our understanding and teaching of best practices regarding biosafety.

From a security perspective, many people are concerned that it is now possible to directly construct harmful pathogens from DNA sequence information. This seems to me a real but remote possibility, and is likely best addressed by improvements in our capacity to respond to emerging infectious diseases, natural or otherwise, and to our public health systems. The more pressing security concern is to ensure that the tools and policies defining the future of biotechnology do not directly or inadvertently lead to a remilitarization of biology by nations.

Finally, synthetic biology advances are challenging the existing application of property rights in biotechnology. Stated plainly, as our capacity to engineer biology increases, so does the number and combinations of uses of genetic functions that will be deployed. Such novel uses and combinations are typically protected via patents. However, via synthetic biology, we are already experiencing situations in which the cost and time required to use a patent-based approach does not match the scale or pace of work. This emerging situation is likely to be exacerbated via an increased capacity to "compile" genetic material from sequences distributed via computer networks, in a fashion that should be familiar to anyone who has used or uses Napster, the Pirate Bay, or iTunes. Our capacity to explore and craft any improved ownership, sharing, and innovation frameworks underlying the future of biotechnology will have direct impacts on the development, application, and ultimate utility and acceptance of synthetic biology.

Thank you.

References

"The 2003 DARPA ISAT Synthetic Biology Study." MIT DSpace Archive, http://hdl.handle.net/1721.1/38455

"Synthetic biology: caught between property rights, the public domain, and the commons." Rai A, Boyle J., *PLoS Biol.* 2007 Mar;5(3):e58

"Foundations for engineering biology." Endy D., *Nature.* 2005 Nov 24;438(7067):449- 53.

"DNA synthesis and biological security." Bügl H, Danner JP, Molinari RJ, Mulligan JT, Park HO, Reichert B, Roth DA, Wagner R, Budowle B, Scripp RM, Smith JA, Steele SJ, Church G, Endy D. *Nat Biotechnol.* 2007 Jun;25(6):627-9.

"Synthetic Genomics: Options for Governance." Garfinkel M; Endy D; Epstein G; Friedman R, MIT DSpace Archive, http://hdl.handle.net/ 1721.1/39141

"The Imperative of Synthetic Biology: A Proposed National Research Initiative." Lazowksa E, Endy D, MIT DSpace Archive, http://hdl.handle. net/1721.1/43950

In: Synthetic Biology
Editors: Roger E. Hamilton and Laura M. Morales
ISBN: 978-1-62257-207-6

Chapter 6

TESTIMONY OF GREGORY E. KAEBNICK, THE HASTINGS CENTER. HEARING ON "EFFECTS OF DEVELOPMENTS IN SYNTHETIC GENOMICS"*

Mr. Chairman, Ranking Member Barton, and Distinguished Members of the Committee, thank you for inviting me to these hearings and for bringing attention to the ethical and social issues raised by this significant new field.

I am a research scholar at The Hastings Center, an independent, nonpartisan, and nonprofit research institute that has been studying ethical issues in medicine, health policy, medical research, and biotechnology since 1969. I am also editor of one of our journals, the *Hastings Center Report*. We are now in the second year of a two-year project, funded by the Alfred P. Sloan Foundation, to investigate the ethical issues of synthetic biology.

My goal this morning is chiefly to set synthetic biology within a widely accepted framework for thinking about the ethics of new technologies. I will also comment briefly on the appropriate governmental response to the field.

The ethical issues raised by synthetic biology are familiar themes in an ongoing conversation this nation has been having about biotechnologies for several decades. Moreover, I am pleased that the conversation is continuing, not only in this panel, but also with President Obama's Presidential Commission for the Study of Bioethical Issues, which will be holding hearings on synthetic biology this summer and delivering a report to the president later this year.

The concerns fall into two general categories. One has to do with whether the creation of synthetic organisms is a good or a bad thing in and of itself, aside from the consequences. These are thought of as intrinsic concerns.

Many people had similar intrinsic concerns about reproductive cloning, for example; they just felt it was wrong to do, regardless of benefits. Another has to do with potential consequences—that is, with risks and benefits. The distinction between these categories can be difficult to maintain in practice, but it provides a useful organizational structure.

* This is an edited, reformatted and augmented version of testimony given on May 27, 2010 before the House Committee on Energy and Commerce.

1. Intrinsic Concerns

I will start with the more philosophical, maybe more baffling, kind of concern—the intrinsic concerns. They are an appropriate place to start because the work just published by researchers at Synthetic Genomics, Inc., has been billed as advancing our understanding of these issues in addition to making a scientific advance. This announcement is not the first time we have had a debate about whether biotechnology challenges deeply held views about the status of life and the power that biotechnology and medicine give us over it. There was a similar debate about gene transfer research in the 1970s and 1980s, about cloning and stem cell research in the 1990s, and—particularly in the last decade but also earlier—about various tools for enhancing human beings. They have been addressed by the President's Commission for the Study of Ethical Problems in Medicine and Biomedical and Behavioral Research in 1983, by President Clinton's National Bioethics Advisory Council, and by President Bush's President's Council on Bioethics. These concerns are related to even older concerns in medicine about decisions to withhold or withdraw medical treatment at the end of life. The fact that we have had this debate before speaks to its importance. I believe the intrinsic concerns deserve respect, and with some kinds of biotechnology I think they are very important, but for synthetic biology, I do not think they provide a basis for decisions about governance.

A. Religious or Metaphysical Concerns

The classic concern about synthetic biology is that it puts human beings in a role properly held by God—that scientists who do it are "playing God," as people say. Some may also believe that life is sacred, and that scientists are violating its sacredness. Prince Charles had this in mind in a famous polemic some years ago when he lamented that biotechnology was leading to "the industrialisation of Life."

To object to synthetic biology along these lines is to see a serious moral mistake in it. This kind of objection may be grounded in deeply held beliefs about God's goals in creating the world and the proper role of human beings within God's plan. But these views would belong to particular faiths— not everybody would share them. Moreover, there is a range of opinions even within religious traditions about what human beings may and may not do. Some people celebrate human creativity and science. They may see science as a gift from God that God intends human beings to develop and use.

The announcement that Synthetic Genomics, Inc., has created a synthetic cell appears to some to disprove the view that life is sacred, but I do not agree. Arguably, what has been created is a synthetic genome, not a completely synthetic cell. Even if scientists manage to create a fully synthetic cell, however, people who believe that life is sacred, that it is something more than interacting chemicals, could continue to defend that belief. A similar question arises about the existence of souls in cloned people: If people have souls, then surely they would have souls even if they were created in the laboratory by means of cloning techniques. By the same reasoning, if microbial life is more than a combination of chemicals, then even microbial life created in the laboratory would be more than just chemicals. In general, beliefs about the sacredness of life are not undermined by science. Moreover, even the creation of a truly synthetic cell would still start with existing materials. It would not be the kind of creating with which God is credited, which is creating something from nothing—creation ex nihilo.

B. Concerns that Synthetic Biology Will Undermine Morally Significant Concepts

A related but different kind of concern is that synthetic biology will simply undermine our shared understanding of important moral concepts. For example, perhaps it will lead us to think that life does not have the specialness we have often found in it, or that we humans are more powerful than we have thought in the past. This kind of concern can be expressed without talking about God's plan.

Synthetic biology need not change our understanding of the value of life, however. The fact that living things are created naturally, rather than by people, would be only one reason for seeing them as valuable, and we could continue to see them as valuable when they are created by people. Further, in its current form, synthetic biology is almost exclusively about engineering single-celled organisms, which may be less troubling to people than engineering more complex organisms. If the work is contained within the laboratory and the factory, then it might not end up broadly changing humans' views of the value of life.

Also, of course, the fact that the work challenges our ideas may not really be a moral problem. It would not be the first time that science has challenged our views of life or our place in the cosmos, and we have weathered these challenges in the past.

C. Concerns about the Human Relationship to Nature

Another way of saying that there's something intrinsically troubling about synthetic biology, again without necessarily talking about the possibility that people are treading on God's turf, is to see it as a kind of environmentalist concern. Many environmentalists want to do more than make the environment good for humans; they also want to save nature from humans—they want to save endangered species, wildernesses, "wild rivers," old-growth forests, and mountains, canyons, and caves, for example. We should approach the natural world, many feel, with a kind of reverence or gratitude, and some worry that synthetic biology—perhaps along with many other kinds of biotechnology—does not square with this value.

Of course, human beings have been altering nature throughout human history. They have been altering ecosystems, affecting the survival of species, affecting the evolution of species, and even creating new species. Most agricultural crop species, for example, are dramatically different from their ancestral forebears. The issue, then, is where to draw the line. Even people who want to preserve nature accept that there is a balance to be struck between saving trees and harvesting them for wood. There might also be a balance when it comes to biotechnology. The misgiving is that synthetic biology goes too far—it takes human control over nature to the ultimate level, where we are not merely altering existing life forms but creating new forms.

Another environmentalist perspective, however, is that synthetic biology could be developed so that it is beneficial to the environment. Synthetic Genomics, Inc. recently contracted with Exxon Mobil to engineer algae that produce gasoline in ways that not only eliminate some of the usual environmental costs of producing and transporting fuel but simultaneously absorb large amounts of carbon dioxide, thereby offsetting some of the environmental costs of burning fuel (no matter how it is produced). If that could be achieved, many who feel deeply that we should tread more lightly on the natural world might well find synthetic biology attractive. In order to achieve this benefit, however, we must be confident that synthetic organisms will not escape into the environment and cause harms there.

CONCERNS INVOLVING CONSEQUENCES

The second category of moral concerns is about consequences—that is, risks and benefits. The promise of synthetic biology includes, for example, better ways of producing medicine, environmentally friendlier ways of producing fuel and other substances, and remediation of past environmental damage. These are not morally trivial considerations. There are also, however, morally serious risks. These, too, fall into three categories.

Concerns about Social Justice

Synthetic biology is sometimes heralded as the start of a new industrial age. Not only will it lead to new products, but it will lead to new modes of production and distribution; instead of pumping oil out of the ground and shipping it around the world, we might be able to produce it from algae in places closer to where it will be used. Inevitably, then, it would have all sorts of large-scale economic and social consequences, some of which could be harmful and unjust. Some commentators hold, for example, that if synthetic biology generates effective ways of producing biofuels from feedstocks such as sugar cane, then farmland in poor countries would be converted from food production to sugar cane production. Another set of concerns arises over the intellectual property rights in synthetic biology. If synthetic biology is the beginning of a new industrial age, and a handful of companies received patents giving them broad control over it, the results could be unjust.

Surely we ought to avoid these consequences. It is my belief that we can do so without avoiding the technology. Also, traditional industrial methods themselves seem to be leading to disastrous longterm social consequences; if so, synthetic biology might provide a way toward better social outcomes.

Concerns about Biosafety

Another concern is about biosafety—about mechanisms for containing and controlling synthetic organisms, both during research and development and in industrial applications. The concern is that organisms will escape, turn out to have properties, at least in their new environment, different from what was intended and predicted, or maybe mutate to acquire them, and then pose a threat to public health, agriculture, or the environment. Alternatively, some of their genes might be transferred to other, wild microbes, producing wild microbes with new properties.

Controlling this risk means controlling the organisms—trying to prevent industrial or laboratory accidents, and then trying to make sure that, when organisms do escape, they are not dangerous. Many synthetic biologists argue that an organism that devotes most of its energy to producing jet fuel or medicine, that is greatly simplified (so that it lacks the genetic complexity and therefore the adaptability of a wild form), and that is designed to work in a controlled, contained environment, will simply be too weak to survive in the wild. For added assurance, perhaps engineering them with failsafe mechanisms will *ensure* that they are incapable of surviving in the wild.

Concerns about Deliberate Misuse

I once heard a well-respected microbiologist say that he was very enthusiastic about synthetic biology, and that the only thing that worries him is the possibility of catastrophe. The kind of thing that worries him is certainly possible. The 1918 flu virus has been recreated in the laboratory. In 2002, a scientist in New York stitched together stretches of nucleotides to produce a string of DNA that was equivalent to RNA polio virus and eventually produced the RNA virus using the DNA string. More recently, the SARS virus was also created in the laboratory. Eventually, it will almost certainly be possible to recreate bacterial pathogens like smallpox. We might also be able to enhance these pathogens. Some work in Australia on mousepox suggests ways of making smallpox more potent, for example. In theory, entirely new pathogens could be created. Pathogens that target crops or livestock are also possible.

Controlling this risk means controlling the people and companies who have access to DNA synthesis or the tools they could use to synthesize DNA themselves. There are some reasons to think that the worst will never actually happen. To be wielded effectively, destructive synthetic organisms would also have to be weaponized; for example, methods must be found to disperse pathogens in forms that will lead to epidemic infection in the target population while sparing one's own population. Arguably, terrorists have better forms of attacking their enemies than with bioweapons, which are still comparatively hard to make and are very hard to control. However, our policy should amount to more than hoping for the best.

Governance

In assessing these risks and establishing oversight over synthetic biology, we do not start from square one. There is an existing framework of laws and regulations, put into action by various agencies and oversight bodies, that will apply to R&D and to different applications. The NIH is extending its guidelines for research on genetic engineering to ensure that they are applicable to research on synthetic biology. These Guidelines are enforced by the NIH's Recombinant DNA Advisory Committee and a network of Institutional Biosafety Committees at research institutions receiving federal funding. Many applications would fall under the purview of various federal laws and the agencies that enforce them. For example, a plan to release synthetic organisms into the sea to produce nutrients that would help rebuild ocean food chains would have to pass muster with the EPA. The USDA and FDA also have regulatory authority over applications. The FBI and the NIH's National Science Advisory Board for Biosecurity are formulating policy to regulate the sale of synthetic DNA sequences that might pose a threat to biosecurity.

At the same time, the current regulatory framework may need to be augmented. First, there are questions about whether the existing laws leave gaps. Research conducted by entirely privately funded laboratory might not covered by the NIH's Guidelines, for example. Field testing of a synthetic organism—that is, release into the environment as part of basic research—might not be covered by the existing regulations of the EPA or the USDA. Questions about the adequacy of existing regulations are even more pointed when it comes to concerns about biosecurity, particularly if or when powerful benchtop synthesizers are available in every lab.

The other big question is whether the regulatory bodies' ability to do risk assessment of synthetic biology is adequate. Synthetic biology differs from older forms of genetic engineering in that a synthetic organism could combine DNA sequences found originally in many different organisms, or might even contain entirely novel genetic code. The eventual behavior of these organisms in new environments, should they accidentally end up in one, may therefore be hard to predict.

The synthetic biologists' goal of simplicity is crucial. One of the themes of traditional biology is that living things are usually more complex than they first appear. We should not assume at the outset that synthetic organisms will shed the unpredictability inherent to life. Life tends to find a way. As a starting assumption, we should expect that artificial life will try to find a way as well.

Another difficulty in assessing concerns about both biosafety and deliberate misuse is that, if the field evolves so that important and even innovative work could be done in small, private labs, even in homes, then it could be very difficult to monitor and regulate. The threats of biosafety and deliberate misuse would have to be taken yet more seriously.

Concluding Comments

I take seriously concerns that synthetic biology is bad in and of itself, and I believe that they warrant a thorough public airing, but I do not believe that they provide a good basis for restraining the technology, at least if we can be confident that the organisms will not lead to environmental damage. Better yet would be to get out in front of the technology and ensure that it benefits the environment. Possibly, some potential applications of synthetic biology are more troubling than others and should be treated differently.

Ultimately, I think the field should be assessed on its possible outcomes. At the moment, we do not understand the possible outcomes well enough. We need, I believe:

- more study of the emergence, plausibility, and impact of potential risks;
- a strategy for studying the risks that is multidisciplinary, rather than one conducted entirely within the field;
- a strategy that is grounded in good science rather than sheer speculation, yet flexible enough to look for the unexpected; and
- an analysis of whether our current regulatory framework is adequate to deal with these risks and how the framework should be augmented.

Different kinds of applications pose different risks and may call for different responses. Microbes intended for release into the environment, for example, would pose a different set of concerns than microbes designed to be kept in specialized, contained settings. Overall, however, while the risks of synthetic biology are too significant to leave the field alone, its potential benefits are too great to call for a general moratorium.

Thank you for this opportunity to share my thoughts.

In: Synthetic Biology
Editors: Roger E. Hamilton and Laura M. Morales
ISBN: 978-1-62257-207-6

Chapter 7

TESTIMONY OF DR. ANTHONY S. FAUCI, DIRECTOR, NATIONAL INSTITUTE OF ALLERGY AND INFECTIOUS DISEASES, NATIONAL INSTITUTES OF HEALTH. HEARING ON "EFFECTS OF DEVELOPMENTS IN SYNTHETIC GENOMICS"*

ADVANCES IN SYNTHETIC BIOLOGY: SIGNIFICANCE AND IMPLICATIONS

Mr. Chairman and members of the Committee, thank you for the opportunity today to discuss the recent advance in synthetic biology made by Dr. J. Craig Venter and his colleagues at the J. Craig Venter Institute (JCVI), the potential practical applications of this advance, and the broader implications of synthetic biology.

I am Dr. Anthony Fauci, Director of the National Institute of Allergy and Infectious Diseases (NIAID), the lead component of the National Institutes of Health (NIH), an agency of the Department of Health and Human Services (HHS), for research relating to infectious diseases, including research on the genomics of infectious microbes.

NIAID Research

NIAID supports research related to the basic understanding, treatment and prevention of infectious, immune-mediated and allergic diseases that threaten millions of human lives. NIAID-supported studies include basic research, such as microbial biology and physiology; applied research, including the development of medical diagnostics, therapeutics and vaccines; and clinical trials to evaluate experimental drugs and vaccines. We also conduct and sponsor research to understand the genomes of disease-causing microbes. A genome is the

* This is an edited, reformatted and augmented version of testimony given on May 27, 2010 before the House Committee on Energy and Commerce.

complete set of DNA (or in some cases, RNA) that contains the genes and instructions—a blueprint—for the maintenance, growth, and reproduction of an organism.

Research fields such as genomics are creating a wealth of information about infectious diseases. Using advanced technologies, researchers are developing a clearer understanding of infectious microbes, the mechanisms by which they cause disease, and the host immune responses necessary to prevent and control an infection.

NIAID has made significant investments in genomic-related activities that provide comprehensive genomic, functional genomic, bioinformatics, and proteomic resources to the scientific community for basic and applied research to rapidly address the Institute's mission and meet the public health preparedness needs of the United States and the world. NIAID-supported researchers have sequenced the complete genomes of hundreds of disease-causing organisms, defining the genetic blueprint for pathogens responsible for malaria, tuberculosis, and seasonal and pandemic influenza, among others. Data generated through NIAID-supported initiatives are rapidly made available to the research community. NIAID genomic programs not only provide the scientific community with valuable research resources, but also have enhanced NIAID research efforts in a number of areas including studies of the mechanisms by which microbes cause disease, and the development of drugs, vaccines, and diagnostics.

Although NIAID did not fund the work by the JCVI that we are discussing here today, we are supporting a number of investigators who are conducting research applying recombinant DNA technologies, genomics, and other related disciplines to study infectious diseases.

Research on genomics and other advanced technologies supported by the NIH and other federal departments and agencies—as well as private entities such as JCVI —will provide the knowledge that will allow further advances in the field of synthetic biology.

Synthetic Biology and Its Practical Applications

"Synthetic biology" can be defined a number of ways. Generally, it is considered to be the use of molecular biological techniques and chemical synthesis to mimic and even redesign natural biological systems. Advances in recombinant DNA technology and genomics, in a sense, represent the early stages along the continuum of synthetic biology and have laid the groundwork for the next frontiers of synthetic biology, which I will discuss later in my testimony.

In the early 1970s, the advent of genetic engineering using recombinant DNA technology revolutionized molecular biology. These technologies refer to techniques by which DNA molecules that code for a protein of interest are either cut out of another genome or, as technical advances occurred, are synthesized using the blueprint of a known genetic sequence. Then, by a variety of enzymatic techniques, these genetic sequences or genes are transferred into another organism. This modified organism then uses its own genetic capabilities together with the inserted gene to produce the protein of interest. These techniques of genetic engineering have been invaluable in biological and medical research, and have led to important, practical medical applications. In 1982, the Food and Drug Administration (FDA) approved the first medicine made by recombinant DNA technology—human insulin produced from a recombinant strain of the bacteria *Escherichia coli*. In 1986, the first recombinant vaccine was approved by the FDA—a vaccine against hepatitis B virus.

As you know, last week Dr. Venter and colleagues announced that they were able to chemically synthesize the entire genome of *Mycoplasma mycoides* based on the known sequence of the microbe, replace the DNA from the bacterium *Mycoplasma capricolum* with this synthetic genome, and produce functioning bacterial cells that mimicked *M. mycoides*. This research is an important technical breakthrough in synthetic biology and our efforts to engineer and potentially synthesize novel microbes that are able to benefit humans and the environment. The potential practical applications of this advance are broad. Certainly, we hope that synthetic organisms might one day be used to create new biofuels. Organisms might be engineered to degrade waste and byproducts that are detrimental to the environment. Scientists might one day be able to create organisms that have a positive impact on agriculture and food production. And, there also are possible medical applications, including the production of biological products and vaccines.

While this is no doubt an important technical breakthrough and a leap forward for the field of synthetic biology, there is still much work to be done in this field. The researchers at JCVI took the known sequence of *M. mycoides*—its genetic blueprint—and were able to mimic it synthetically, but this effort was more complex and challenging than anticipated, occupying many talented scientists for more than a decade at the cost tens of millions of dollars. One task ahead is the creation of a new blueprint—something that does not yet exist—that will perform the task that scientists ask it to do: sequester carbon dioxide, produce fuel, clean up waste, etc. The creation of a completely novel blueprint from scratch will be extraordinarily challenging because scientists are only beginning to understand all of the intricate circuits involved to put together such a blueprint.

Synthetic Biology and Its Broader Implications

As is the case with many of the genomic technologies that have been developed over the last several decades, synthetic biology technology potentially could be used to engineer microbes that are beneficial, but also to create microbes that may be harmful to humans and the environment. Such technologies that have both beneficial and potentially harmful applications are commonly referred to as "dual use".

While the advance made by JCVI scientists potentially could be used by those who intend to do harm, we also must recognize that this was not a simple experiment; it was an extraordinarily complex project that took many years, people, and millions of dollars to complete. While there certainly is a chance that the technology developed by JCVI researchers might be used for nefarious purposes by those with extensive resources, it is important to point out that similar, albeit simpler techniques, are in widespread usage and are an integral and vital tool in life science research and science education, including high school through post-graduate curricula. We also must keep in mind that nature itself is already an expert at creating microbes that can cause great harm to humans. This recent advance in synthetic biology docs not necessarily bring us closer to harm's way than existing technologies or nature itself.

Dual-use technologies, including synthetic biology, have been the subject of active and ongoing discussions in the scientific community for many years, and Dr. Venter and his colleagues have been active and invaluable contributors to this dialogue. This dialogue involving a substantial number of scientists has taken place in national and international scientific bodies such as the U.S. National Academy of Sciences and the Royal Society, the United Kingdom's national academy of science. Advisory bodies to the federal government, such as the NIH Recombinant DNA Advisory Committee (RAC) and the National Science Advisory Board for Biosecurity (NSABB) also have played a major role in this discourse.

The RAC was established in 1974 in response to public concerns about the safety of manipulating genetic material through the use of recombinant DNA techniques. While the membership and responsibilities of the RAC have evolved with technology over the years, it continues to serve the NIH, as well as the scientific community and lay public, as a critically important forum for open, public deliberation on the scientific, ethical, and legal issues raised by recombinant DNA technology and its basic and clinical research applications.

The RAC first issued the NIH Guidelines for Research Involving Recombinant DNA Molecules (found at http://oba.od.nih.gov/rdna/ nih_ guidelines_oba.html) in 1976. While compliance with the NIH Guidelines is mandatory for investigators at institutions that receive NIH funding for recombinant DNA research, the NIH Guidelines have become a universal standard for safe scientific practice in this area of research. Other federal agencies, such as U.S. Department of Agriculture, the Department of Energy, and the Department of Veterans Affairs, have made compliance with the NIH Guidelines a term and condition of their awards. The NIH Guidelines also are followed voluntarily by many companies and other institutions not otherwise subject to their requirements. At its quarterly meetings, the RAC discusses research proposals that raise novel or particularly important scientific, safety, or ethical considerations. By helping to foster awareness and understanding of these matters among scientists, the RAC has fostered a cultured of responsibility among the genomic sciences community.

In addition to the RAC, NIH also manages the NSABB, which was established in 2004 to advise the Federal government on strategies to minimize the risks and harm that could result from the malevolent use of information from legitimate research, i.e., dual-use research. The NSABB advises HHS on the efficient and effective oversight of federally conducted or supported dual-use biological research, taking into consideration national security concerns and the needs of the research community. The NSABB also provides advice on the interpretation and application of federal guidelines on dual-use research in instances where a research institution seeks additional advice.

In addition to the dual-use implications, the recent advance in synthetic biology has broadened the range of organisms that may be developed, including those with entirely novel functions. As such, this advance raises broader societal and ethical concerns about this and future advances in this field. As such, the President has directed the Presidential Commission for the Study of Bioethical Issues to undertake, as its first order of business, a study of the implications of this scientific milestone, as well as other advances that may lie ahead in this field of research. The President has asked that the Commission consult with a broad range of constituencies and provide a report of its finding and recommendations within six months.

As I have described here, the Federal government has a number of existing committees and advisory bodies that have been discussing and will continue to discuss the risks and benefits related to this advance. The President has acted swiftly to ensure that more comprehensive discussions of its implications occur outside of the scientific community. In addition, the federal government will review its existing authorities to ensure that the current legal, regulatory, and oversight framework is sufficient to mitigate the risks associated with synthetic biology.

Conclusion

As discussed above, the advance announced by Dr. Venter and his colleagues is an important step forward for the field of synthetic biology. While it is important to ensure that we proceed cautiously with this technology and protect the public against its potential misuse, we also must take care to avoid any hasty response that would harm our scientific enterprise and hamper scientific progress. It is important that we do not act rashly and place undue restrictions on our best and brightest scientists that would prevent the United States from developing and utilizing this technology effectively and responsibly for the good of mankind in addition to competing effectively with other countries who will surely adopt these techniques.

INDEX

C

D

E

H

I

J

K

L

M

N

O

P

Q

R

S

T

U

V

W

Y